STUDENT SOLUTIONS MANUAL

to accompany

APPLIED CALCULUS
Fifth Edition

Deborah Hughes-Hallett
University of Arizona

Andrew M. Gleason
Harvard University

Patti Frazer Lock
St. Lawrence University

Daniel E. Flath
Macalester College

et al.

Prepared by
Elliot J. Marks

WILEY

COVER PHOTO © Patrick Zephyr / Patrick Zephyr Nature Photography

To order books or for customer service please, call 1-800-CALL WILEY (225-5945).

This material is based upon work supported by the National Science Foundation under Grant No. DUE-9352905. Opinions expressed are those of the authors and not necessarily those of the Foundation.

ISBN-13 978-1-118-71499-7

10 9 8 7 6 5 4 3 2 1

Table of Contents

CHAPTER ONE

Solutions for Section 1.1

1. **(a)** The story in (a) matches Graph (IV), in which the person forgot her books and had to return home.
   (b) The story in (b) matches Graph (II), the flat tire story. Note the long period of time during which the distance from home did not change (the horizontal part).
   (c) The story in (c) matches Graph (III), in which the person started calmly but sped up later.
      The first graph (I) does not match any of the given stories. In this picture, the person keeps going away from home, but his speed decreases as time passes. So a story for this might be: *I started walking to school at a good pace, but since I stayed up all night studying calculus, I got more and more tired the farther I walked.*

5. The noise level is going down as distance goes up. A possible graph is shown in Figure 1.1. The graph is decreasing.

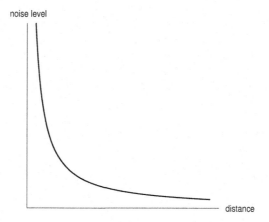

noise level

distance

Figure 1.1

9. **(a)** At $p = 0$, we see $r = 8$. At $p = 3$, we see $r = 7$.
   (b) When $p = 2$, we see $r = 10$. Thus, $f(2) = 10$.

13. Looking at the graph, we see that the point on the graph with an x-coordinate of 5 has a y-coordinate of 2. Thus

$$f(5) = 2.$$

17. The year 2008 was 0 years before 2008 so 2008 corresponds to $t = 0$. Thus, an expression that represents the statement is:

$$f(0) \text{ meters.}$$

21. **(a)** $f(30) = 10$ means that the value of f at $t = 30$ was 10. In other words, the temperature at time $t = 30$ minutes was $10°C$. So, 30 minutes after the object was placed outside, it had cooled to 10 °C.
   (b) The intercept a measures the value of $f(t)$ when $t = 0$. In other words, when the object was initially put outside, it had a temperature of $a°C$. The intercept b measures the value of t when $f(t) = 0$. In other words, at time b the object's temperature is 0 °C.

25. **(a)** The original deposit is the balance, B, when $t = 0$, which is the vertical intercept. The original deposit was $1000.
   (b) It appears that $f(10) \approx 2200$. The balance in the account after 10 years is about $2200.
   (c) When $B = 5000$, it appears that $t \approx 20$. It takes about 20 years for the balance in the account to reach $5000.

29. One possible graph is shown in Figure 1.2. Since the patient has none of the drug in the body before the injection, the vertical intercept is 0. The peak concentration is labeled on the concentration (vertical) axis and the time until peak concentration is labeled on the time (horizontal) axis. See Figure 1.2.

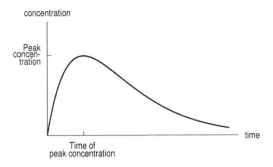

Figure 1.2

33. (a) (I) The incidence of cancer increase with age, but the rate of increase slows down slightly. The graph is nearly linear. This type of cancer is closely related to the aging process.

    (II) In this case a peak is reached at about age 55, after which the incidence decreases.

    (III) This type of cancer has an increased incidence until the age of about 48, then a slight decrease, followed by a gradual increase.

    (IV) In this case the incidence rises steeply until the age of 30, after which it levels out completely.

    (V) This type of cancer is relatively frequent in young children, and its incidence increases gradually from about the age of 20.

    (VI) This type of cancer is not age-related – all age-groups are equally vulnerable, although the overall incidence is low (assuming each graph has the same vertical scale).

(b) Graph (V) shows a relatively high incidence rate for children. Leukemia behaves in this way.

(c) Graph (III) could represent cancer in women with menopause as a significant factor. Breast cancer is a possibility here.

(d) Graph (I) shows a cancer which might be caused by toxins building up in the body. Lung cancer is a good example of this.

37. See Figure 1.3.

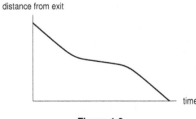

Figure 1.3

Solutions for Section 1.2

1. The slope is $(3 - 2)/(2 - 0) = 1/2$. So the equation of the line is $y = (1/2)x + 2$.

5. Rewriting the equation as

$$y = -\frac{12}{7}x + \frac{2}{7}$$

shows that the line has slope $-12/7$ and vertical intercept $2/7$.   s

9. The slope is positive for lines l_1 and l_2 and negative for lines l_3 and l_4. The y-intercept is positive for lines l_1 and l_3 and negative for lines l_2 and l_4. Thus, the lines match up as follows:

(a) l_1

(b) l_3

(c) l_2

(d) l_4

13. (a) The vertical intercept appears to be about 300 miles. When this trip started (at $t = 0$) the person was 300 miles from home.

(b) The slope appears to be about $(550 - 300)/5 = 50$ miles per hour. This person is moving away from home at a rate of about 50 miles per hour.

(c) We have $D = 300 + 50t$.

17. (a) The slope is 1.8 billion dollars per year. McDonald's revenue is increasing at a rate of 1.8 billion dollars per year.

(b) The vertical intercept is 19.1 billion dollars. In 2005, McDonald's revenue was 19.1 billion dollars.

(c) Substituting $t = 10$, we have $R = 19.1 + 1.8 \cdot 10 = 37.1$ billion dollars.

(d) We substitute $R = 35$ and solve for t:

$$19.1 + 1.8t = 35$$
$$1.8t = 15.9$$
$$t = 8.83 \text{ years.}$$

The function predicts that annual revenue will hit 35 billion dollars in 2013.

21. (a) We want $P = b + mt$, where t is in years since 1975. The slope is

$$\text{Slope } = m = \frac{\Delta P}{\Delta t} = \frac{2048 - 1241}{30 - 0} = 26.9.$$

Since $P = 1241$ when $t = 0$, the vertical intercept is 1241 so

$$P = 1241 + 26.9t.$$

(b) The slope tells us that world grain production has been increasing at a rate of 26.9 million tons per year.

(c) The vertical intercept tells us that world grain production was 1241 million tons in 1975 (when $t = 0$.)

(d) We substitute $t = 40$ to find $P = 1241 + 26.9 \cdot 40 = 2317$ million tons of grain.

(e) Substituting $P = 2500$ and solving for t, we have

$$2500 = 1241 + 26.9t$$
$$1259 = 26.9t$$
$$t = 46.803.$$

Grain production is predicted to reach 2500 million tons during the year 2021.

25. (a) We are given two points: $N = 34$ when $l = 11$ and $N = 26$ when $l = 44$. We use these two points to find the slope. Since $N = f(l)$, we know the slope is $\Delta N / \Delta l$:

$$m = \frac{\Delta N}{\Delta l} = \frac{26 - 34}{44 - 11} = -0.2424.$$

We use the point $l = 11, N = 34$ and the slope $m = -0.2424$ to find the vertical intercept:

$$N = b + ml$$
$$34 = b - 0.2424 \cdot 11$$
$$34 = b - 2.67$$
$$b = 36.67.$$

The equation of the line is

$$N = 36.67 - 0.2424l.$$

(b) The slope is -0.2424 species per degree latitude. In other words, the number of coastal dune plant species decreases by about 0.2424 for every additional degree South in latitude. The vertical intercept is 36.67 species, which represents the number of coastal dune plant species at the equator, if the model (and Australia) extended that far.

(c) See Figure 1.4.

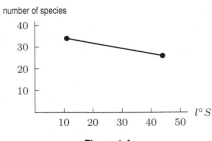

number of species

Figure 1.4

29. If f is the age of the female, and m the age of the male, then if they have the same MHR,

$$226 - f = 220 - m,$$

so

$$f - m = 6.$$

Thus the female is 6 years older than the male.

33. (a) We have $y = 3.5$ when $u = 0$, which occurs when the unemployment rate does not change. When the unemployment rate is constant, Okun'a law states that US production increases by 3.5% annually.

(b) When unemployment rises from 5% to 8% we have $u = 3$ and therefore

$$y = 3.5 - 2u = 3.5 - 2 \cdot 3 = -2.5.$$

National production for the year decreases by 2.5%.

(c) If annual production does not change, then $y = 0$. Hence $0 = 3.5 - 2u$ and so $u = 1.75$. There is no change in annual production if the unemployment rate goes up 1.75%.

(d) The coefficient -2 is the slope $\Delta y / \Delta u$. Every 1% increase in the unemployment rate during the course of a year results in an additional 2% decrease in annual production for the year.

Solutions for Section 1.3

1. The graph shows a concave down function.

5. As t increases w decreases, so the function is decreasing. The rate at which w is decreasing is itself decreasing: as t goes from 0 to 4, w decreases by 42, but as t goes from 4 to 8, w decreases by 36. Thus, the function is concave up.

9. The average rate of change R between $x = -2$ and $x = 1$ is:

$$R = \frac{f(1) - f(-2)}{1 - (-2)} = \frac{3(1)^2 + 4 - (3(-2)^2 + 4)}{1 + 2} = \frac{7 - 16}{3} = -3.$$

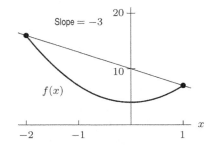

13. (a) The average rate of change is the change in attendance divided by the change in time. Between 2003 and 2007,

$$\text{Average rate of change} = \frac{22.26 - 21.64}{2007 - 2003} = 0.155 \text{ million people per year}$$
$$= 155{,}000 \text{ people per year.}$$

(b) For each of the years from 2003–2007, the annual increase in the number of games was:

$$2003 \text{ to } 2004 : 21.71 - 21.64 = 0.07$$
$$2004 \text{ to } 2005 : 0.08$$
$$2005 \text{ to } 2006 : 0.41$$
$$2006 \text{ to } 2007 : 0.06.$$

(c) We average the four yearly rates of change:

$$\text{Average of the four figures in part (b)} = \frac{0.07 + 0.08 + 0.41 + 0.06}{4}$$
$$= \frac{0.62}{4} = 0.155, \text{ which is the same as part (a).}$$

17. (a) We have

$$\text{Average rate of change of } P = \frac{6.45 - 5.68}{2005 - 1995} = 0.077 \text{ billion people per year.}$$
$$\text{Average rate of change of } A = \frac{45.9 - 36.1}{2005 - 1995} = 0.980 \text{ million cars per year.}$$
$$\text{Average rate of change of } C = \frac{2168 - 91}{2005 - 1995} = 207.7 \text{ million subscribers per year.}$$

(b) (i) The number of people is increasing faster since the population is increasing at 77 million per year and car production is increasing at less than 1 million per year.

(ii) The number of cell phone subscribers is increasing faster since the population is increasing at 77 million per year and the number of cell phone subscribers is increasing at about 208 million per year.

21. (a) The average rate of change R is the difference in amounts divided by the change in time.

$$R = \frac{719 - 802}{2010 - 2003}$$
$$= \frac{-83}{7}$$
$$\approx -11.86 \quad \text{million pounds/yr.}$$

This means that in the years between 2003 and 2010, the production of tobacco decreased at a rate of approximately 11.86 million pounds per year.

(b) To have a positive rate of change, the production has to increase during one of these 1 year intervals. Looking at the data, we can see that between 2003 and 2004, production of tobacco increased by 80 million pounds. Thus, the average rate of change is positive between 2003 and 2004.

Likewise, the annual average rate of change is positive between 2005 and 2009, but the large decline in 2005 accounts for the negative rate of change over the period 2003 to 2010.

25. (a) The change is given by

$$\text{Change between 2003 and 2008} = \text{Revenues in 2008} - \text{Revenues in 2003}$$
$$= 149.0 - 184.0$$
$$= -35 \text{ billion dollars.}$$

(b) The rate of change is given by

$$\begin{aligned}
\text{Average rate of change} \atop \text{between 2003 and 2008} &= \frac{\text{Change in revenues}}{\text{Change in time}} \\
&= \frac{\text{Revenues in 2008} - \text{Revenues in 2003}}{2008 - 2003} \\
&= \frac{149.0 - 184.0}{2008 - 2003} \\
&= -7 \text{ billion dollars per year.}
\end{aligned}$$

This means that General Motors' revenues decreased on average by 7 billion dollars per year between 2003 and 2008.

(c) From 2003 to 2008 there were two one-year time intervals during which the average rate of change in revenues was negative: -24.5 billion dollars per year between 2006 and 2007, and -32.1 billion dollars per year between 2007 and 2008.

29. Between 1804 and 1927, the world's population increased 1 billion people in 123 years, for an average rate of change of $1/123$ billion people per year. We convert this to people per minute:

$$\frac{1,000,000,000}{123} \text{ people/year} \cdot \frac{1}{60 \cdot 24 \cdot 365} \text{ years/minute} = 15.468 \text{ people/minute.}$$

Between 1804 and 1927, the population of the world increased at an average rate of 15.468 people per minute. Similarly, we find the following:

Between 1927 and 1960, the increase was 57.654 people per minute.

Between 1960 and 1974, the increase was 135.899 people per minute.

Between 1974 and 1987, the increase was 146.353 people per minute.

Between 1987 and 1999, the increase was 158.549 people per minute.

33. (a)   (i) $f(1985) = 13$

   (ii) $f(1990) = 99$

(b) The average yearly increase is the rate of change.

$$\text{Yearly increase} = \frac{f(1990) - f(1985)}{1990 - 1985} = \frac{99 - 13}{5} = 17.2 \text{ billionaires per year.}$$

(c) Since we assume the rate of increase remains constant, we use a linear function with slope 17.2 billionaires per year. The equation is

$$f(t) = b + 17.2t$$

where $f(1985) = 13$, so

$$13 = b + 17.2(1985)$$
$$b = -34,129.$$

Thus, $f(t) = 17.2t - 34,129$.

37. Between 0 and 21 years,

$$\text{Average rate of change} = -\frac{9}{21} = -0.429 \text{ beats per minute per year,}$$

whereas between 0 and 33 years,

$$\text{Average rate of change} = -\frac{26}{33} = -0.788 \text{ beats per minute per year.}$$

Because the average rate of change is negative, a decreasing function is suggested. Also, as age increases, the average rate of change decreases, suggesting the graph of the function is concave down. (Since the average rate of change is negative and increasing in absolute value, this rate is decreasing.)

41. We have

$$\text{Relative change} = \frac{0.05 - 0.3}{0.3} = -0.833.$$

The quantity W decreases by 83.3%.

45. The changes are 5 students for the small class and 20 students for the larger class. The relative changes are

$$\frac{5}{5} = 100\% \quad \text{and} \quad \frac{20}{30} = 66.667\%.$$

The relative change for the small class is bigger than the relative change for the large class, but the extra 20 students added to a class of 30 has a potentially larger effect on the class.

49. We have

$$\text{Relative change} = \frac{\text{Change in cost}}{\text{Initial cost}} = \frac{46 - 45}{45} = 0.022.$$

The cost to mail a letter increased by 2.2%.

53. (a) The largest time interval was 2005–2007 since the percentage growth rate decreased from -1.9 in 2005 to -45.4 in 2007. This means that from 2005 to 2007 the US consumption of hydroelectric power shrunk relatively more with each successive year.

(b) The largest time interval was 2004–2007 since the percentage growth rates were negative for each of these four consecutive years. This means that the amount of hydroelectric power consumed by the US industrial sector steadily decreased during the four year span from 2004 to 2007, then increased in 2008.

Solutions for Section 1.4

1. See Figure 1.5

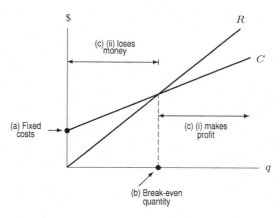

Figure 1.5

5. The 5.7 represents the fixed cost of $5.7 million.

The 0.002 represents the variable cost per unit in millions of dollars. Thus, each additional unit costs an additional $0.002 million $= \$2000$ to produce.

9. The 5500 tells us the quantity that would be demanded if the price were $0. The 100 tells us the change in the quantity demanded if the price increases by $1. Thus, a price increase of $1 causes demand to drop by 100 units.

13. The cost function $C(q) = b + mq$ satisfies $C(0) = 5000$, so $b = 5000$, and $MC = m = 15$. So

$$C(q) = 5000 + 15q.$$

The revenue function is $R(q) = 60q$, so the profit function is

$$\pi(q) = R(q) - C(q) = 60q - 5000 - 15q = 45q - 5000.$$

17. (a) We know that the fixed cost of the first price list is $100 and the variable cost is $0.03. Thus, the cost of making q copies under the first option is

$$C_1(q) = 100 + 0.03q.$$

We know that the fixed cost of the second price list is $200 and the variable cost is $0.02. Thus, the cost of making q copies under the second option is

$$C_2(q) = 200 + 0.02q.$$

(b) At 5000 copies, the first price list gives the cost

$$C_1(5000) = 100 + 5000(0.03) = 100 + 150 = \$250.$$

At 5000 copies, the second price list gives the cost

$$C_2(5000) = 200 + 5000(0.02) = 200 + 100 = \$300.$$

Thus, for 5000 copies, the first price list is cheaper.

(c) We are asked to find the point q at which

$$C_1(q) = C_2(q).$$

Solving we get

$$C_1(q) = C_2(q)$$
$$100 + 0.03q = 200 + 0.02q$$
$$100 = 0.01q$$
$$q = 10,000.$$

Thus, if one needs to make ten thousand copies, the cost under both price lists will be the same.

21. (a) The cost function is of the form

$$C(q) = b + m \cdot q$$

where m is the variable cost and b is the fixed cost. Since the variable cost is $20 and the fixed cost is $650,000, we get

$$C(q) = 650,000 + 20q.$$

The revenue function is of the form

$$R(q) = pq$$

where p is the price that the company is charging the buyer for one pair. In our case the company charges $70 a pair so we get

$$R(q) = 70q.$$

The profit function is the difference between revenue and cost, so

$$\pi(q) = R(q) - C(q) = 70q - (650,000 + 20q) = 70q - 650,000 - 20q = 50q - 650,000.$$

(b) Marginal cost is $20 per pair. Marginal revenue is $70 per pair. Marginal profit is $50 per pair.

(c) We are asked for the number of pairs of shoes that need to be produced and sold so that the profit is larger than zero. That is, we are trying to find q such that

$$\pi(q) > 0.$$

Solving we get

$$\pi(q) > 0$$
$$50q - 650,000 > 0$$
$$50q > 650,000$$
$$q > 13,000.$$

Thus, if the company produces and sells more than 13,000 pairs of shoes, it will make a profit.

25. We know that at the point where the price is $1 per scoop the quantity must be 240. Thus we can fill in the graph as follows:

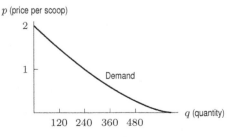

Figure 1.6

(a) Looking at Figure 1.6 we see that when the price per scoop is half a dollar, the quantity given by the demand curve is roughly 360 scoops.

(b) Looking at Figure 1.6 we see that when the price per scoop is $1.50, the quantity given by the demand curve is roughly 120 scoops.

29. (a) We know that the cost function will be of the form

$$C = mq + b$$

where m is the variable cost and b is the fixed cost. In this case this gives

$$C = 5q + 7000.$$

We know that the revenue function is of the form

$$R = pq$$

where p is the price per shirt. Thus in this case we have

$$R = 12q.$$

(b) We are given

$$q = 2000 - 40p.$$

We are asked to find the demand when the price is \$12. Plugging in $p = 12$ we get

$$q = 2000 - 40(12) = 2000 - 480 = 1520.$$

Given this demand we know that the cost of producing $q = 1520$ shirts is

$$C = 5(1520) + 7000 = 7600 + 7000 = \$14,600.$$

The revenue from selling $q = 1520$ shirts is

$$R = 12(1520) = \$18,240.$$

Thus the profit is

$$\pi(12) = R - C$$

or in other words

$$\pi(12) = 18,240 - 14,600 = \$3640.$$

(c) Since we know that

$$q = 2000 - 40p,$$
$$C = 5q + 7000,$$

and

$$R = pq,$$

we can write

$$C = 5q + 7000 = 5(2000 - 40p) + 7000 = 10,000 - 200p + 7000 = 17,000 - 200p$$

and

$$R = pq = p(2000 - 40p) = 2000p - 40p^2.$$

We also know that the profit is the difference between the revenue and the cost so

$$\pi(p) = R - C = 2000p - 40p^2 - (17,000 - 200p) = -40p^2 + 2200p - 17,000.$$

(d) Looking at Figure 1.7 we see that the maximum profit occurs when the company charges about \$27.50 per shirt. At this price, the profit is about \$13,250.

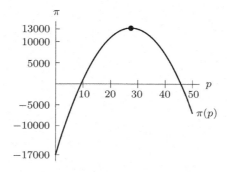

Figure 1.7

33. One possible supply curve is shown in Figure 1.8. Many other answers are possible.

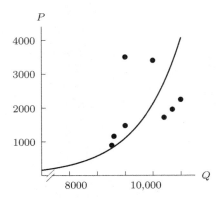

Figure 1.8

37. (a) See Figure 1.9.
   (b) If the slope of the supply curve increases then the supply curve will intersect the demand curve sooner, resulting in a higher equilibrium price p_1 and lower equilibrium quantity q_1. Intuitively, this makes sense since if the slope of the supply curve increases. The amount produced at a given price decreases. See Figure 1.10.

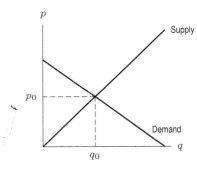

Figure 1.9

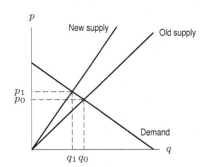

Figure 1.10

   (c) When the slope of the demand curve becomes more negative, the demand function will decrease more rapidly and will intersect the supply curve at a lower value of q_1. This will also result in a lower value of p_1 and so the equilibrium price p_1 and equilibrium quantity q_1 will decrease. This follows our intuition, since if demand for a product lessens, the price and quantity purchased of the product will go down. See Figure 1.11.

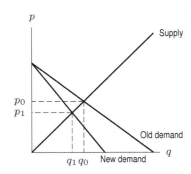

Figure 1.11

41. Before the tax is imposed, the equilibrium is found by solving the equations $q = 0.5p - 25$ and $q = 165 - 0.5p$. Setting the values of q equal, we have

$$0.5p - 25 = 165 - 0.5p$$
$$p = 190 \text{ dollars}$$

Substituting into one of the equation for q, we find $q = 0.5(190) - 25 = 70$ units. Thus, the pre-tax equilibrium is

$$p = \$190, \quad q = 70 \text{ units.}$$

The original supply equation, $q = 0.5p - 25$, tells us that

$$\text{Quantity supplied} = 0.5 \left(\begin{array}{c} \text{Amount per unit} \\ \text{received by suppliers} \end{array} \right) - 25.$$

When the tax is imposed, the suppliers receive only $p - 8$ dollars per unit because \$8 goes to the government as taxes. Thus, the new supply curve is

$$q = 0.5(p - 8) - 25 = 0.5p - 29.$$

The demand curve is still

$$q = 165 - 0.5p.$$

To find the equilibrium, we solve the equations $q = 0.5p - 29$ and $q = 165 - 0.5p$. Setting the values of q equal, we have

$$0.5q - 29 = 165 - 0.5p$$
$$q = 194 \text{ dollars}$$

Substituting into one of the equations for q, we find that $q = 0.5(194) - 29 = 68$ units. Thus, the post-tax equilibrium is

$$p = \$194, q = 68 \text{ units.}$$

Solutions for Section 1.5

1. (a) Town (i) has the largest percent growth rate, at 12%.
   (b) Town (ii) has the largest initial population, at 1000.
   (c) Yes, town (iv) is decreasing in size, since the decay factor is 0.9, which is less than 1.

5. This looks like an exponential decay function $y = Ca^t$. The y-intercept is 30 so we have $y = 30a^t$. We use the point $(25, 6)$ to find a:

$$y = 30a^t$$
$$6 = 30a^{25}$$
$$0.2 = a^{25}$$
$$a = (0.2)^{1/25} = 0.94.$$

The formula is $y = 30(0.94)^t$.

9. (a) This is a linear function with slope -2 grams per day and intercept 30 grams. The function is $Q = 30 - 2t$, and the graph is shown in Figure 1.12.

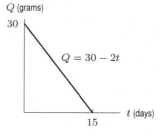

Figure 1.12

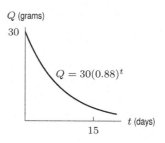

Figure 1.13

(b) Since the quantity is decreasing by a constant percent change, this is an exponential function with base $1 - 0.12 = 0.88$. The function is $Q = 30(0.88)^t$, and the graph is shown in Figure 1.13.

13. Since the decay rate is 2.9%, the base of the exponential function is $1 - 0.029 = 0.971$. If P represents the percent of forested land in 1980 remaining t years after 1980, when $t = 0$ we have $P = 100$. The formula is $P = 100(0.971)^t$. The percent remaining in 2015 will be $P = 100(0.971)^{35} = 35.7$. About 35.7% of the land will be covered in forests 35 years later if the rate continues.

17. Each increase of 1 in t seems to cause $g(t)$ to decrease by a factor of 0.8, so we expect an exponential function with base 0.8. To make our solution agree with the data at $t = 0$, we need a coefficient of 5.50, so our completed equation is

$$g(t) = 5.50(0.8)^t.$$

21. (a) We divide the two equations to solve for a:

$$\frac{P_0 a^4}{P_0 a^3} = \frac{18}{20}$$
$$a = 0.9$$

Now that we know the value of a, we can use either of the equations to solve for P_0. Using the first equation, we have:

$$P_0 a^4 = 18$$
$$P_0(0.9^4) = 18$$
$$P_0 = \frac{18}{0.9^4} = 27.435$$

We see that $a = 0.9$ and $P_0 = 27.435$.

(b) The initial quantity is 27.435 and the quantity is decaying at a rate of 10% per unit time.

25. Use an exponential function of the from $P = P_0 a^t$ for the number of passengers. At $t = 0$, the number of passengers is $P_0 = 190,205$, so

$$P = 190,205a^t.$$

After $t = 5$ years, the number of passengers is 174,989, so

$$174,989 = 190,205a^5$$
$$a^5 = \frac{174,989}{190,205} = 0.920002$$
$$a = (0.920002)^{1/5} = 0.983.$$

Then

$$P = 190,205(0.932)^t,$$

which means the annual percentage decrease over this period is $1 - 0.983 = 0.017$ or 1.7%.

29. (a) A linear function must change by exactly the same amount whenever x changes by some fixed quantity. While $h(x)$ decreases by 3 whenever x increases by 1, $f(x)$ and $g(x)$ fail this test, since both change by different amounts between $x = -2$ and $x = -1$ and between $x = -1$ and $x = 0$. So the only possible linear function is $h(x)$, so it will be given by a formula of the type: $h(x) = mx + b$. As noted, $m = -3$. Since the y-intercept of h is 31, the formula for $h(x)$ is $h(x) = 31 - 3x$.

(b) An exponential function must grow by exactly the same factor whenever x changes by some fixed quantity. Here, $g(x)$ increases by a factor of 1.5 whenever x increases by 1. Since the y-intercept of $g(x)$ is 36, $g(x)$ has the formula $g(x) = 36(1.5)^x$. The other two functions are not exponential; $h(x)$ is not because it is a linear function, and $f(x)$ is not because it both increases and decreases.

33. (a) Using $Q = Q_0(1 - r)^t$ for loss, we have

$$Q = 10,000(1 - 0.1)^{10} = 10,000(0.9)^{10} = 3486.78.$$

The investment was worth \$3486.78 after 10 years.

(b) Measuring time from the moment at which the stock begins to gain value and letting $Q_0 = 3486.78$, the value after t years is

$$Q = 3486.78(1 + 0.1)^t = 3486.78(1.1)^t.$$

We can estimate the value of t when $Q = 10,000$ by tracing along a graph of Q, giving $t \approx 11$. It will take about 11 years to get the investment back to \$10,000.

37. (a) Since the annual growth factor from 2005 to 2006 was $1 + 1.866 = 2.866$ and $91(1 + 1.866) = 260.806$, the US consumed approximately 261 million gallons of biodiesel in 2006. Since the annual growth factor from 2006 to 2007 was $1 + 0.372 = 1.372$ and $261(1 + 0.372) = 358.092$, the US consumed about 358 million gallons of biodiesel in 2007.

(b) Completing the table of annual consumption of biodiesel and plotting the data gives Figure 1.14.

Year	2005	2006	2007	2008	2009
Consumption of biodiesel (mn gal)	91	261	358	316	339

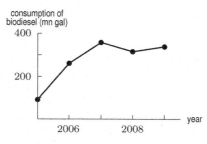

Figure 1.14

41. (a) The US consumption of wind power energy increased by at least 40% in 2006 and in 2008, relative to the previous year. In 2006 consumption increased by just under 50% over consumption in 2005, and in 2008 consumption increased by about 60% over consumption in 2007. Consumption did not decrease during the time period shown because all the annual percent growth values are positive, indicating a steady increase in the US consumption of wind power energy between 2005 and 2009.

(b) Yes. From 2006 to 2007 consumption increased by about 30%, which means $x(1 + 0.30)$ units of wind power energy were consumed in 2007 if x had been consumed in 2006. Similarly,

$$(x(1 + 0.30))(1 + 0.60)$$

units of wind power energy were consumed in 2008 if x had been consumed in 2006 (because consumption increased by about 60% from 2007 to 2008). Since

$$(x(1 + 0.30))(1 + 0.60) = x(2.08) = x(1 + 1.08),$$

the percent growth in wind power consumption was about 108%, or just over 100%, in 2008 relative to consumption in 2006.

Solutions for Section 1.6

1. Taking natural logs of both sides we get

$$\ln 10 = \ln(2^t).$$

This gives

$$t \ln 2 = \ln 10$$

or in other words

$$t = \frac{\ln 10}{\ln 2} \approx 3.3219$$

5. Taking natural logs of both sides we get

$$\ln(e^t) = \ln 10$$

which gives

$$t = \ln 10 \approx 2.3026.$$

9. Taking natural logs of both sides we get

$$\ln(e^{3t}) = \ln 100.$$

This gives

$$3t = \ln 100$$

or in other words

$$t = \frac{\ln 100}{3} \approx 1.535.$$

13. Dividing both sides by P we get

$$\frac{B}{P} = e^{rt}.$$

Taking the natural log of both sides gives

$$\ln(e^{rt}) = \ln\left(\frac{B}{P}\right).$$

This gives

$$rt = \ln\left(\frac{B}{P}\right) = \ln B - \ln P.$$

Dividing by r gives

$$t = \frac{\ln B - \ln P}{r}.$$

17. Initial quantity $= 5$; growth rate $= 0.07 = 7\%$.

21. Since $e^{0.25t} = \left(e^{0.25}\right)^t \approx (1.2840)^t$, we have $P = 15(1.2840)^t$. This is exponential growth since 0.25 is positive. We can also see that this is growth because $1.2840 > 1$.

25. Since we want $(1.5)^t = e^{kt} = (e^k)^t$, so $1.5 = e^k$, and $k = \ln 1.5 = 0.4055$. Thus, $P = 15e^{0.4055t}$. Since 0.4055 is positive, this is exponential growth.

29. We use the information to create two equations: $P_0 e^{3k} = 140$ and $P_0 e^{1k} = 100$.

   (a) We divide the two equations to solve for k:

$$\frac{P_0 e^{3k}}{P_0 e^k} = \frac{140}{100}$$
$$e^{2k} = 1.4$$
$$2k = \ln 1.4$$
$$k = \frac{\ln 1.4}{2} = 0.168$$

Now that we know the value of k, we can use either of the equations to solve for P_0. Using the first equation, we have:

$$P_0 e^{3k} = 140$$
$$P_0 e^{3(0.168)} = 140$$
$$P_0 = \frac{140}{e^{0.504}} = 84.575$$

We see that $a = 0.168$ and $P_0 = 84.575$.

   (b) The initial quantity is 84.575 and the quantity is growing at a continuous rate of 16.8% per unit time.

33. We find a with $a^t = e^{0.08t}$. Thus, $a = e^{0.08} = 1.0833$. The corresponding annual percent growth rate is 8.33%.

37. **(a)** $P = 5.4(1.034)^t$

   (b) Since $P = 5.4e^{kt} = 5.4(1.034)^t$, we have

$$e^{kt} = (1.034)^t$$
$$kt = t\ln(1.034)$$
$$k = 0.0334.$$

Thus, $P = 5.4e^{0.0334t}$.

   (c) The annual growth rate is 3.4%, while the continuous growth rate is 3.3%. Thus the growth rates are not equal, though for small growth rates (such as these), they are close. The annual growth rate is larger.

41. (a) Since the percent rate of growth is constant and given as a continuous rate, we use the exponential function $S = 6.1e^{0.042t}$.

(b) In 2015, we have $t = 4$ and $S = 6.1e^{0.042\cdot4} = 7.21$ billion dollars.

(c) Using the graph in Figure 1.15, we see that $S = 8$ at approximately $t = 6.5$. Solving $6.1e^{0.042t} = 8$, we have

$$e^{0.042t} = \frac{8}{6.1}$$

$$t = \frac{\ln(8/6.1)}{0.042} = 6.456 \text{ years.}$$

In other words, assuming this growth rate continues, annual net sales are predicted to pass 8 billion dollars midway through 2017.

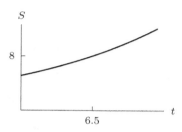

Figure 1.15

45. If C_0 is the concentration of NO_2 on the road, then the concentration x meters from the road is

$$C = C_0 e^{-0.0254x}.$$

We want to find the value of x making $C = C_0/2$, that is,

$$C_0 e^{-0.0254x} = \frac{C_0}{2}.$$

Dividing by C_0 and then taking natural logs yields

$$\ln\left(e^{-0.254x}\right) = -0.0254x = \ln\left(\frac{1}{2}\right) = -0.6931,$$

so

$$x = 27 \text{ meters.}$$

At 27 meters from the road the concentration of NO_2 in the air is half the concentration on the road.

Solutions for Section 1.7

1. In both cases the initial deposit was $20. Compounding continuously earns more interest than compounding annually at the same interest rate. Therefore, curve A corresponds to the account which compounds interest continuously and curve B corresponds to the account which compounds interest annually. We know that this is the case because curve A is higher than curve B over the interval, implying that bank account A is growing faster, and thus is earning more money over the same time period.

5. We use the equation $B = Pe^{rt}$. We want to have a balance of $B = \$20,000$ in $t = 6$ years, with an annual interest rate of 10%.

$$20,000 = Pe^{(0.1)6}$$

$$P = \frac{20,000}{e^{0.6}}$$

$$\approx \$10,976.23.$$

9. Every 2 hours, the amount of nicotine remaining is reduced by $1/2$. Thus, after 4 hours, the amount is $1/2$ of the amount present after 2 hours.

Table 1.1

t (hours)	0	2	4	6	8	10
Nicotine (mg)	0.4	0.2	0.1	0.05	0.025	0.0125

From the table it appears that it will take just over 6 hours for the amount of nicotine to reduce to 0.04 mg.

13. (a) Since the initial amount of caffeine is 100 mg and the exponential decay rate is -0.17, we have $A = 100e^{-0.17t}$.
(b) See Figure 1.16. We estimate the half-life by estimating t when the caffeine is reduced by half (so $A = 50$); this occurs at approximately $t = 4$ hours.

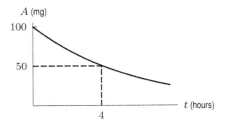

Figure 1.16

(c) We want to find the value of t when $A = 50$:

$$50 = 100e^{-0.17t}$$
$$0.5 = e^{-0.17t}$$
$$\ln 0.5 = -0.17t$$
$$t = 4.077.$$

The half-life of caffeine is about 4.077 hours. This agrees with what we saw in Figure 1.16.

17. If the rate of increase is $r\%$ then each year the output will increase by $1 + r/100$ so that after five years the level of output is

$$20{,}000\left(1 + \frac{r}{100}\right)^5.$$

To achieve a final output of 30,000, the value of r must satisfy the equation

$$20{,}000\left(1 + \frac{r}{100}\right)^5 = 30{,}000.$$

Dividing both sides by 20,000 gives

$$\left(1 + \frac{r}{100}\right)^5 = 1.5,$$

so $(1 + \frac{r}{100}) = 1.5^{1/5} = 1.0845$. Solving for r gives an annual growth rate of 8.45%.

21. (a) At $t = 0$, the population is 1000. The population doubles (reaches 2000) at about $t = 4$, so the population doubled in about 4 years.
(b) At $t = 3$, the population is about 1700. The population reaches 3400 at about $t = 7$. The population doubled in about 4 years.
(c) No matter when you start, the population doubles in 4 years.

25. Since the amount of strontium-90 remaining halves every 29 years, we can solve for the decay constant;

$$0.5P_0 = P_0 e^{-29k}$$
$$k = \frac{\ln(1/2)}{-29}.$$

Knowing this, we can look for the time t in which $P = 0.10P_0$, or

$$0.10P_0 = P_0 e^{\ln(0.5)t/29}$$
$$t = \frac{29\ln(0.10)}{\ln(0.5)} = 96.336 \text{ years.}$$

29. We assume exponential decay and solve for k using the half-life:

$$e^{-k(5730)} = 0.5 \quad \text{so} \quad k = 1.21 \cdot 10^{-4}.$$

Now find t, the age of the painting:

$$e^{-1.21 \cdot 10^{-4} t} = 0.995, \quad \text{so} \quad t = \frac{\ln 0.995}{-1.21 \cdot 10^{-4}} = 41.43 \text{ years.}$$

Since Vermeer died in 1675, the painting is a fake.

33. We have

$$\text{Future value} = 10,000 e^{0.03 \cdot 8} = \$12,712.49.$$

37. (a) The total present value for each of the two choices are in the following table. Choice 1 is the preferred choice since it has the larger present value.

Choice 1			Choice 2		
Year	Payment	Present value	Payment	Present value	
0	1500	1500	1900	1900	
1	3000	$3000/(1.05) = 2857.14$	2500	$2500/(1.05) = 2380.95$	
	Total	4357.14	Total	4280.95	

(b) The difference between the choices is an extra \$400 now (\$1900 in Choice 2 instead of \$1500 in Choice 1) versus an extra \$500 one year from now (\$3000 in Choice 1 instead of \$2500 in Choice 2). Since $400 \times 1.25 = 500$, Choice 2 is better at interest rates above 25%.

41. (a) Option 1 is the best option since money received now can earn interest.

(b) In Option 1, the entire \$2000 earns 5% interest for 1 year, so we have:

$$\text{Future value of Option 1} = 2000 e^{0.05 \cdot 1} = \$2102.54.$$

In Option 2, the payment in one year does not earn interest, but the payment made now earns 5% interest for one year. We have

$$\text{Future value of Option 2} = 1000 + 1000 e^{0.05 \cdot 1} = 1000 + 1051.27 = \$2051.27.$$

Since Option 3 is paid all in the future, we have

$$\text{Future value of Option 3} = \$2000.$$

As we expected, Option 1 has the highest future value.

(c) In Option 1, the entire \$2000 is paid now, so we have:

$$\text{Present value of Option 1} = \$2000.$$

In Option 2, the payment now has a present value of \$1000 but the payment in one year has a lower present value. We have

$$\text{Present value of Option 2} = 1000 + 1000 e^{-0.05 \cdot 1} = 1000 + 951.23 = \$1951.23.$$

Since Option 3 is paid all in the future, we have

$$\text{Present value of Option 3} = 2000 e^{-0.05 \cdot 1} = \$1902.46.$$

Again, we see that Option 1 has the highest value.
Alternatively, we could have computed the present values directly from the future values found in part (b).

45. The following table contains the present value of each of the payments, though it does not take into account the resale value if you buy the machine.

Buy			Lease	
Year	Payment	Present Value	Payment	Present Value
0	12000	12000	2650	2650
1	580	$580/(1.0775) = 538.28$	2650	$2650/(1.0775) = 2459.40$
2	464	$464(1.0775)^2 = 399.65$	2650	$2650/(1.0775)^2 = 2282.50$
3	290	$290/(1.0775)^3 = 231.82$	2650	$2650/(1.0775)^3 = 2118.33$
	Total	13,169.75	Total	9510.23

Now we consider the $5000 resale.

$$\text{Present value of resale} = \frac{5000}{(1.0775)^3} = 3996.85.$$

The net present value associated with buying the machine is the present value of the payments minus the present value of the resale price, which is

$$\text{Present value of buying} = 13{,}169.75 - 3996.85 = 9172.90$$

Since the present value of the expenses associated with buying ($9172.90) is smaller than the present value of leasing ($9510.23), you should buy the machine.

Solutions for Section 1.8

1. (a) We have $f(g(x)) = f(3x + 2) = 5(3x + 2) - 1 = 15x + 9$.
(b) We have $g(f(x)) = g(5x - 1) = 3(5x - 1) + 2 = 15x - 1$.
(c) We have $f(f(x)) = f(5x - 1) = 5(5x - 1) - 1 = 25x - 6$.

5. (a) $g(2 + h) = (2 + h)^2 + 2(2 + h) + 3 = 4 + 4h + h^2 + 4 + 2h + 3 = h^2 + 6h + 11$.
(b) $g(2) = 2^2 + 2(2) + 3 = 4 + 4 + 3 = 11$, which agrees with what we get by substituting $h = 0$ into (a).
(c) $g(2 + h) - g(2) = (h^2 + 6h + 11) - (11) = h^2 + 6h$.

9. (a) $f(g(1)) = f(1^2) = f(1) = e^1 = e$
(b) $g(f(1)) = g(e^1) = g(e) = e^2$
(c) $f(g(x)) = f(x^2) = e^{x^2}$
(d) $g(f(x)) = g(e^x) = (e^x)^2 = e^{2x}$
(e) $f(t)g(t) = e^t t^2$

13.

(a)

x	$f(x) + 3$
0	13
1	9
2	6
3	7
4	10
5	14

(b)

x	$f(x - 2)$
2	10
3	6
4	3
5	4
6	7
7	11

(c)

x	$5g(x)$
0	10
1	15
2	25
3	40
4	60
5	75

(d)

x	$-f(x) + 2$
0	-8
1	-4
2	-1
3	-2
4	-5
5	-9

(e)

x	$g(x - 3)$
3	2
4	3
5	5
6	8
7	12
8	15

(f)

x	$f(x) + g(x)$
0	12
1	9
2	8
3	12
4	19
5	26

17. $m(z+h) - m(z) = (z+h)^2 - z^2 = 2zh + h^2$.

21. We see in the graph of $f(x)$ that $f(1) = 5$ so we have $g(f(1)) = g(5) = 6$.

25. We see in the graph of $g(x)$ that $g(2) = 3$ so we have $g(g(2)) = g(3) = 4$.

29. We see in the graph of $g(x)$ that $g(3) \approx -2.5$ so we have $f(g(3)) = f(-2.5) \approx 0$.

33. We have approximately $v(40) = 15$ and $u(15) = 18$ so $u(v(40)) = 18$.

37. The graph of $y = f(x)$ is shifted vertically upward by 1 unit. See Figure 1.17.

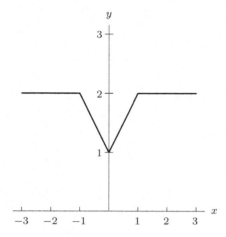

Figure 1.17: Graph of $y = f(x) + 1$

41. The graph of $y = f(x)$ is reflected over the x-axis and shifted vertically up by 3 units. See Figure 1.18.

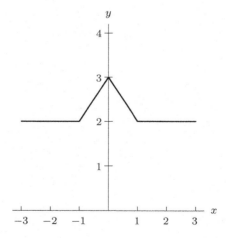

Figure 1.18: Graph of $y = -f(x) + 3$

45.

(a)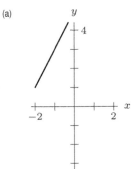

(b)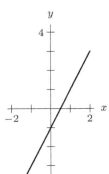

(c)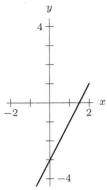

(d)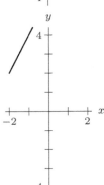

(e)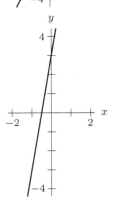

(f)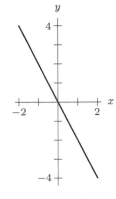

49. See Figure 1.19.

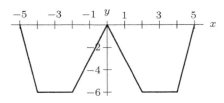

Figure 1.19: Graph of $y = -3f(x)$

53. **(a)** The equation is $y = 2x^2 + 1$. Note that its graph is narrower than the graph of $y = x^2$ which appears in gray. See Figure 1.20.

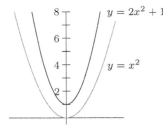

Figure 1.20

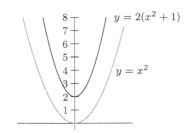

Figure 1.21

(b) $y = 2(x^2 + 1)$ moves the graph up one unit and *then* stretches it by a factor of two. See Figure 1.21.

(c) No, the graphs are not the same. Since $2(x^2 + 1) = (2x^2 + 1) + 1$, the second graph is always one unit higher than the first.

57. The time elapsed is: $f^{-1}(g^{-1}(10{,}000))$ min.

Solutions for Section 1.9

1. $y = (1/5)x$; $k = 1/5$, $p = 1$.

5. Not a power function.

9. Not a power function.

13. For some constant k, we have $S = kh^2$.

17. (a) This is a graph of $y = x^3$ shifted to the right 2 units and up 1 unit. A possible formula is $y = (x - 2)^3 + 1$.
    (b) This is a graph of $y = -x^2$ shifted to the left 3 units and down 2 units. A possible formula is $y = -(x + 3)^2 - 2$.

21. (a) Since daily calorie consumption, C, is proportional to the 0.75 power of weight, W, we have

$$C = kW^{0.75}.$$

    (b) Since the exponent for this power function is positive and less than one, the graph is increasing and concave down. See Figure 1.22.

    (c) We use the information about the human to find the constant of proportionality k. With W in pounds we have

$$1800 = k(150)^{0.75}$$
$$k = \frac{1800}{150^{0.75}} = 42.0.$$

The function is

$$C = 42.0W^{0.75}.$$

For a horse weighing 700 lbs, the daily calorie requirement is

$$C = 42.0 \cdot 700^{0.75} = 5{,}715.745 \text{ calories.}$$

For a rabbit weighing 9 lbs, the daily calorie requirement is

$$C = 42.0 \cdot 9^{0.75} = 218.238 \text{ calories.}$$

    (d) Because the exponent is less than one, the graph is concave down. A mouse has a faster metabolism and needs to consume more calories per pound of weight.

C (Calories)

W (weight)

Figure 1.22

25. (a) T is proportional to the fourth root of B, and so

$$T = k\sqrt[4]{B} = kB^{1/4}.$$

    (b) $148 = k \cdot (5230)^{1/4}$ and so $k = 148/(5230)^{1/4} = 17.4$.
    (c) Since $T = 17.4B^{1/4}$, for a human with $B = 70$ we have

$$T = 17.4(70)^{1/4} = 50.3 \text{ seconds}$$

It takes about 50 seconds for all the blood in the body to circulate and return to the heart.

29. (a) We know that the demand function is of the form

$$q = mp + b$$

where m is the slope and b is the vertical intercept. We know that the slope is

$$m = \frac{q(30) - q(25)}{30 - 25} = \frac{460 - 500}{5} = \frac{-40}{5} = -8.$$

Thus, we get

$$q = -8p + b.$$

Substituting in the point $(30, 460)$ we get

$$460 = -8(30) + b = -240 + b,$$

so that

$$b = 700.$$

Thus, the demand function is

$$q = -8p + 700.$$

(b) We know that the revenue is given by

$$R = pq$$

where p is the price and q is the demand. Thus, we get

$$R = p(-8p + 700) = -8p^2 + 700p.$$

(c) Figure 1.23 shows the revenue as a function of price.

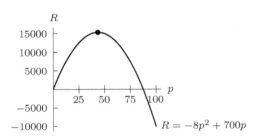

Figure 1.23

Looking at the graph, we see that maximal revenue is attained when the price charged for the product is roughly $44. At this price the revenue is roughly $15,300.

Solutions for Section 1.10

1.

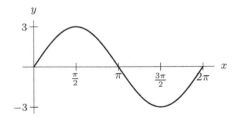

Figure 1.24

See Figure 1.24. The amplitude is 3; the period is 2π.

5.

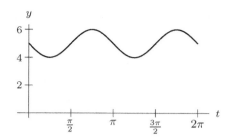

Figure 1.25

See Figure 1.25. The amplitude is 1; the period is π.

9. (a) The function appears to vary between 5 and -5, and so the amplitude is 5.
   (b) The function begins to repeat itself at $x = 8$, and so the period is 8.
   (c) The function is at its highest point at $x = 0$, so we use a cosine function. It is centered at 0 and has amplitude 5, so we have $f(x) = 5\cos(Bx)$. Since the period is 8, we have $8 = 2\pi/B$ and $B = \pi/4$. The formula is

$$f(x) = 5\cos\left(\frac{\pi}{4}x\right).$$

13. The levels of both hormones certainly look periodic. In each case, the period seems to be about 28 days. Estrogen appears to peak around the 12th day. Progesterone appears to peak from the 17th through 21st day. (Note: the days given in the answer are approximate so your answer may differ slightly.)

17. The graph looks like an upside-down sine function with amplitude $(90 - 10)/2 = 40$ and period π. Since the function is oscillating about the line $x = 50$, the equation is

$$x = 50 - 40\sin(2t).$$

21. The graph is a sine curve which has been shifted up by 2, so $f(x) = (\sin x) + 2$.

25. The graph is a cosine curve with period $2\pi/5$ and amplitude 2, so it is given by $f(x) = 2\cos(5x)$.

29. Depth $= 7 + 1.5\sin\left(\frac{\pi}{3}t\right)$

Solutions for Chapter 1 Review

1. Since t represents the number of years since 1970, we see that $f(35)$ represents the population of the city in 2005. In 2005, the city's population was 12 million.

5. The contamination is probably greatest right at the tank, at a depth of 6 meters. Contamination probably goes down as the distance from the tank increases. If we assume that the gas spreads both up and down, the graph might look that the one shown in Figure 1.26.

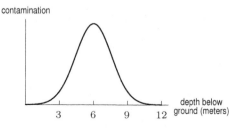

Figure 1.26

9. The equation of the line is of the form $y = b + mx$ where m is the slope given by

$$\text{Slope} = \frac{\text{Rise}}{\text{Run}} = \frac{2-3}{2-(-1)} = -\frac{1}{3}$$

so the line is of the form $y = b - x/3$. For the line to pass through $(-1, 3)$ we need $3 = b + 1/3$ so $b = 8/3$. Therefore, the equation of the line passing through $(-1, 3)$ and $(2, 2)$ is

$$y = \frac{8}{3} - \frac{1}{3}x.$$

Note that when $x = -1$ this gives $y = 3$ and when $x = 2$ it gives $y = 2$ as required.

13. Given that the function is linear, choose any two points, for example $(5.2, 27.8)$ and $(5.3, 29.2)$. Then

$$\text{Slope} = \frac{29.2 - 27.8}{5.3 - 5.2} = \frac{1.4}{0.1} = 14.$$

Using the point-slope formula, with the point $(5.2, 27.8)$, we get the equation

$$y - 27.8 = 14(x - 5.2)$$

which is equivalent to

$$y = 14x - 45.$$

17. The average velocity over a time period is the change in position divided by the change in time. Since the function $s(t)$ gives the position of the particle, we find the values of $s(3) = 12 \cdot 3 - 3^2 = 27$ and $s(1) = 12 \cdot 1 - 1^2 = 11$. Using these values, we find

$$\text{Average velocity} = \frac{\Delta s(t)}{\Delta t} = \frac{s(3) - s(1)}{3 - 1} = \frac{27 - 11}{2} = 8 \text{ mm/sec}.$$

21. The average velocity over a time period is the change in position divided by the change in time. Since the function $s(t)$ gives the position of the particle, we find the values on the graph of $s(3) = 2$ and $s(1) = 3$. Using these values, we find

$$\text{Average velocity} = \frac{\Delta s(t)}{\Delta t} = \frac{s(3) - s(1)}{3 - 1} = \frac{2 - 3}{2} = -\frac{1}{2} \text{ mm/sec}.$$

25. **(a)** This is the graph of a linear function, which increases at a constant rate, and thus corresponds to $k(t)$, which increases by 0.3 over each interval of 1.

(b) This graph is concave down, so it corresponds to a function whose increases are getting smaller, as is the case with $h(t)$, whose increases are 10, 9, 8, 7, and 6.

(c) This graph is concave up, so it corresponds to a function whose increases are getting bigger, as is the case with $g(t)$, whose increases are 1, 2, 3, 4, and 5.

29. Each of these questions can also be answered by considering the slope of the line joining the two relevant points.

(a) The average rate of change is positive if the volume of water is increasing with time and negative if the volume of water is decreasing.

   (i) Since volume is rising from 500 to 1000 from $t = 0$ to $t = 5$, the average rate of change is positive.

   (ii) We can see that the volume at $t = 10$ is greater than the volume at $t = 0$. Thus, the average rate of change is positive.

   (iii) We can see that the volume at $t = 15$ is lower than the volume at $t = 0$. Thus, the average rate of change is negative.

   (iv) We can see that the volume at $t = 20$ is greater than the volume at $t = 0$. Thus, the average rate of change is positive.

(b)   (i) The secant line between $t = 0$ and $t = 5$ is steeper than the secant line between $t = 0$ and $t = 10$, so the slope of the secant line is greater on $0 \le t \le 5$. Since average rate of change is represented graphically by the slope of a secant line, the rate of change in the interval $0 \le t \le 5$ is greater than that in the interval $0 \le t \le 10$.

   (ii) The slope of the secant line between $t = 0$ and $t = 20$ is greater than the slope of the secant line between $t = 0$ and $t = 10$, so the rate of change is larger for $0 \le t \le 20$.

(c) The average rate of change in the interval $0 \le t \le 10$ is about

$$\frac{750 - 500}{10} = \frac{250}{10} = 25 \quad \text{cubic meters per week}$$

This tells us that for the first ten weeks, the volume of water is growing at an average rate of about 25 cubic meters per week.

33. Generally manufacturers will produce more when prices are higher. Therefore, the first curve is a supply curve. Consumers consume less when prices are higher. Therefore, the second curve is a demand curve.

37. Starting with the general exponential equation $y = Ae^{kx}$, we first find that for $(0, 1)$ to be on the graph, we must have $A = 1$. Then to make $(3, 4)$ lie on the graph, we require

$$4 = e^{3k}$$
$$\ln 4 = 3k$$
$$k = \frac{\ln 4}{3} \approx 0.4621.$$

Thus the equation is

$$y = e^{0.4621x}.$$

Alternatively, we can use the form $y = a^x$, in which case we find $y = (1.5874)^x$.

41. We look first at $f(x)$. As x increases by 1, $f(x)$ decreases by 5, then 6, then 7. The rate of change is not constant so the function is not linear. To see if it is exponential, we check ratios of successive terms:

$$\frac{20}{25} = 0.8, \quad \frac{14}{20} = 0.7, \quad \frac{7}{14} = 0.5.$$

Since the ratios are not constant, this function is not exponential either.

What about $g(x)$? As x increases by 1, $g(x)$ decreases by 3.2 each time, so this function is linear with slope -3.2. The vertical intercept (when $x = 0$) is 30.8 so the formula is

$$g(x) = 30.8 - 3.2x.$$

Now consider $h(x)$. As x increases by 1, $h(x)$ decreases by 6000, then 3600, then 2160. The rate of change is not constant so this function is not linear. To see if it is exponential, we check ratios of successive terms:

$$\frac{9000}{15,000} = 0.6, \quad \frac{5400}{9000} = 0.6, \quad \frac{3240}{5400} = 0.6.$$

Since the ratios are constant, this is an exponential function with base 0.6. The initial value (when $x = 0$) is $15,000$ so the formula is

$$h(x) = 15,000(0.6)^x.$$

45. Isolating the exponential term

$$20 = 50(1.04)^x$$
$$\frac{20}{50} = (1.04)^x.$$

Taking logs of both sides

$$\ln \frac{2}{5} = \ln(1.04)^x$$
$$\ln \frac{2}{5} = x \ln(1.04)$$
$$x = \frac{\ln(2/5)}{\ln(1.04)} = -23.4.$$

49. (a) The continuous percent growth rate is 15%.
(b) We want $P_0 a^t = 10e^{0.15t}$, so we have $P_0 = 10$ and $a = e^{0.15} = 1.162$. The corresponding function is

$$P = 10(1.162)^t.$$

(c) Since the base in the answer to part (b) is 1.162, the annual percent growth rate is 16.2%. This annual rate is equivalent to the continuous growth rate of 15%.

(d) When we sketch the graphs of $P = 10e^{0.15t}$ and $P = 10(1.162)^t$ on the same axes, we only see one graph. These two exponential formulas are two ways of representing the same function, so the graphs are the same. See Figure 1.27.

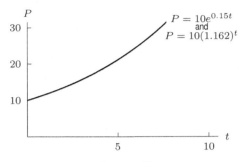

Figure 1.27

53. Given the doubling time of 5 hours, we can solve for the bacteria's growth rate;

$$2P_0 = P_0 e^{k5}$$
$$k = \frac{\ln 2}{5}.$$

So the growth of the bacteria population is given by:

$$P = P_0 e^{\ln(2)t/5}.$$

We want to find t such that

$$3P_0 = P_0 e^{\ln(2)t/5}.$$

Therefore we divide both sides by P_0 and apply ln. We get

$$t = \frac{5\ln(3)}{\ln(2)} = 7.925 \text{ hours.}$$

57. The following table contains the present value of each of the expenses. Since the total present value of the repairs, $255.15, is more than the cost of the service contract, you should buy the service contract.

Present value of repairs		
Year	Repairs	Present Value
1	50	$50/(1.0725) = 46.62$
2	100	$100/(1.0725)^2 = 86.94$
3	150	$150/(1.0725)^3 = 121.59$
	Total	255.15

61. (a) We have $f(g(x)) = f(5x^2) = 2(5x^2) + 3 = 10x^2 + 3$.
   (b) We have $g(f(x)) = g(2x + 3) = 5(2x + 3)^2 = 5(4x^2 + 12x + 9) = 20x^2 + 60x + 45$.
   (c) We have $f(f(x)) = f(2x + 3) = 2(2x + 3) + 3 = 4x + 9$.

65. For $f(x) + 5$, the graph is shifted 5 upward. See Figure 1.28.

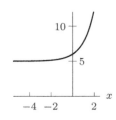

Figure 1.28

69.

(a)

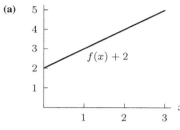

$f(x) + 2$

(b)

$f(x - 1)$

(c)

$3f(x)$

(d)

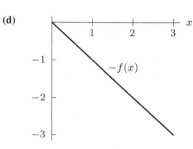

$-f(x)$

73. This graph is the graph of $m(t)$ shifted to the right by 0.5 units and downward by 2.5 units. See Figure 1.29.

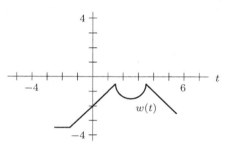

$w(t)$

Figure 1.29

77. If p is proportional to t, then $p = kt$ for some fixed constant k. From the values $t = 10$, $p = 25$, we have $25 = k(10)$, so $k = 2.5$. To see if p is proportional to t, we must see if $p = 2.5t$ gives all the values in the table. However, when we check the values $t = 20, p = 60$, we see that $60 \neq 2.5(20)$. Thus, p is not proportional to t.

81. The period is $2\pi/\pi = 2$, since when t increases from 0 to 2, the value of πt increases from 0 to 2π. The amplitude is 0.1, since the function oscillates between 1.9 and 2.1.

85. This graph is the same as in Problem 19 but shifted up by 2, so it is given by $f(x) = 2\sin\left(\dfrac{x}{4}\right) + 2$.

STRENGTHEN YOUR UNDERSTANDING

1. False, the domain is the set of inputs of a function.

5. True. Plugging in $r = 0$ to find the intercept on the D axis gives $D = 10$.

9. False, if a graph has two vertical intercepts, then it cannot correspond to a function, since the function would have two values at zero.

13. True, any constant function is a linear function with slope zero.

17. False, consider the linear function $y = x - 1$. The slope of this function is 1, and the y-intercept is -1.

21. False. Since the slope of this linear function is negative, the function is decreasing.

25. True, consider $f(x) = \sqrt{x}$.

29. True, for any values $t_1 < t_2$, $\dfrac{s(t_2) - s(t_1)}{t_2 - t_1} = \dfrac{3t_2 + 2 - (3t_1 + 2)}{t_2 - t_1} = \dfrac{3(t_2 - t_1)}{t_2 - t_1} = 3$.

33. True. When quantity increases by 100 from 500 to 600, the cost increases by 15%, which is $0.15 \cdot 1000 = 150$ dollars. Cost goes from \$1000 to \$1150.

37. True, this is the definition of revenue.

41. False. It is possible that supply is greater than demand.

45. False. Imposing a sales tax shifts the curves which can change where they intersect.

49. False, the percent growth rate is 3%.

53. True.

57. True, since $e^0 = 1$.

61. True. We take ln of both sides and bring the exponent t down.

65. True, by the properties of logarithms.

69. False, if an initial investment Q is compounded annually with interest rate r for t years, then the value of the investment is $Q(1 + r)^t$. So, we should have $1000(1.03)^t$ as the value of the account.

73. True, since we have a continuous rate of growth, we use the formula $P_0 e^{kt}$. In our case, $P_0 = 1000$, $k = .03$, and $t = 5$, giving $1000e^{.15}$.

77. True, since $f(x + k)$ is just a shift of the graph of $f(x)$ k units to the left, this shift will not affect the increasing nature of the function f. So, $f(x + k)$ is increasing.

81. False. We have $g(3 + h) = (3 + h)^2 = 9 + 6h + h^2$ so $g(3 + h) \neq 9 + h^2$ for many values of h.

85. False, since $f(x + h) = (x + h)^2 - 1 = x^2 + 2xh + h^2 - 1$, we see that $f(x + h) - f(x) = 2xh + h^2$ which is not equal to h^2.

89. False, this is an exponential function.

93. True.

97. True, for a trig function of the form $\cos(kx) + b$, the period is given by $2\pi/k$. Since $k = 1$ in this case, the period is 2π.

101. False, since the period of $y = \sin(2t)$ is π and the period of $\sin(t)$ is 2π, the period is half.

105. True.

CHAPTER TWO

Solutions for Section 2.1

1. (a) The function $N = f(t)$ is decreasing when $t = 1950$. Therefore, $f'(1950)$ is negative. That means that the number of farms in the US was decreasing in 1950.

(b) The function $N = f(t)$ is decreasing in 1960 as well as in 1980 but it is decreasing faster in 1960 than in 1980. Therefore, $f'(1960)$ is more negative than $f'(1980)$.

5. (a) The average rate of change of a function over an interval is represented graphically as the slope of the secant line to its graph over the interval. See Figure 2.1. Segment AB is the secant line to the graph in the interval from $x = 0$ to $x = 3$ and segment BC is the secant line to the graph in the interval from $x = 3$ to $x = 5$.

We can easily see that slope of $AB >$ slope of BC. Therefore, the average rate of change between $x = 0$ and $x = 3$ is greater than the average rate of change between $x = 3$ and $x = 5$.

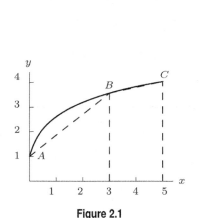

Figure 2.1

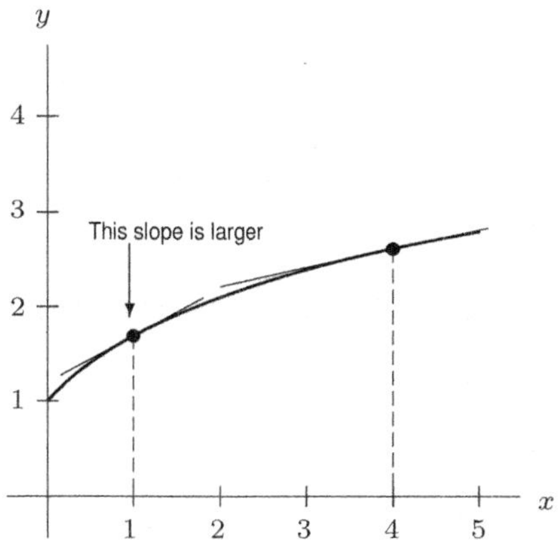

Figure 2.2

(b) We can see from the graph in Figure 2.2 that the function is increasing faster at $x = 1$ than at $x = 4$. Therefore, the instantaneous rate of change at $x = 1$ is greater than the instantaneous rate of change at $x = 4$.

(c) The units of rate of change are obtained by dividing units of cost by units of product: thousands of dollars/kilogram.

9. We use the interval $x = 1$ to $x = 1.01$:

$$g'(1) \approx \frac{f(1.01) - f(1)}{1.01 - 1} = \frac{4^{1.01} - 4^1}{0.01} = \frac{4.05583 - 4}{0.01} = 5.583.$$

For greater accuracy, we can use the smaller interval $x = 1$ to $x = 1.001$:

$$g'(1) \approx \frac{f(1.001) - f(1)}{1.001 - 1} = \frac{4^{1.001} - 4^1}{0.001} = \frac{4.005549 - 4}{0.001} = 5.549.$$

13. Using the interval $1 \le x \le 1.001$, we estimate

$$f'(1) \approx \frac{f(1.001) - f(1)}{0.001} = \frac{3.0033 - 3.0000}{0.001} = 3.3$$

The graph of $f(x) = 3^x$ is concave up so we expect our estimate to be greater than $f'(1)$.

17. The coordinates of A are $(4, 25)$. See Figure 2.3. The coordinates of B and C are obtained using the slope of the tangent line. Since $f'(4) = 1.5$, the slope is 1.5

From A to B, $\Delta x = 0.2$, so $\Delta y = 1.5(0.2) = 0.3$. Thus, at C we have $y = 25 + 0.3 = 25.3$. The coordinates of B are $(4.2, 25.3)$.

From A to C, $\Delta x = -0.1$, so $\Delta y = 1.5(-0.1) = -0.15$. Thus, at C we have $y = 25 - 0.15 = 24.85$. The coordinates of C are $(3.9, 24.85)$.

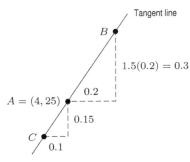

Figure 2.3

21. Using a difference quotient with $h = 0.001$, say, we find

$$f'(1) \approx \frac{1.001 \ln(1.001) - 1 \ln(1)}{1.001 - 1} = 1.0005$$

$$f'(2) \approx \frac{2.001 \ln(2.001) - 2 \ln(2)}{2.001 - 2} = 1.6934$$

The fact that f' is larger at $x = 2$ than at $x = 1$ suggests that f is concave up between $x = 1$ and $x = 2$.

Solutions for Section 2.2

1. The graph is that of the line $y = -2x + 2$. The slope, and hence the derivative, is -2. See Figure 2.4.

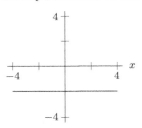

Figure 2.4

5. See Figure 2.5.

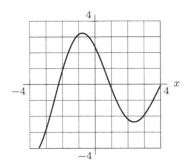

Figure 2.5

9. (a) x_3 (b) x_4 (c) x_5 (d) x_3

13. This is a line with slope 1, so the derivative is the constant function $f'(x) = 1$. The graph is the horizontal line $y = 1$. See Figure 2.6.

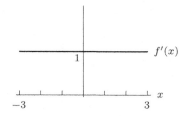

Figure 2.6

17. See Figure 2.7.

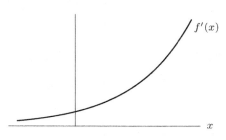

Figure 2.7

21. The function is increasing for $x < -2$ and decreasing for $x > -2$. The corresponding derivative is positive (above the x-axis) for $x < -2$, negative (below the x-axis) for $x > -2$, and zero at $x = -2$. The derivatives in graphs **VI** and **VII** both satisfy these requirements. To decide which is correct, consider what happens as x gets large. The graph of $f(x)$ approaches an asymptote, gets more and more horizontal, and the slope gets closer and closer to zero. The derivative in graph **VI** meets this requirement and is the correct answer.

25. Since $f'(x) > 0$ for $1 < x < 3$, we see that $f(x)$ is increasing on this interval.
Since $f'(x) < 0$ for $x < 1$ and for $x > 3$, we see that $f(x)$ is decreasing on these intervals.
Since $f'(x) = 0$ for $x = 1$ and $x = 3$, the tangent to $f(x)$ will be horizontal at these x's.
One of many possible shapes of $y = f(x)$ is shown in Figure 2.8.

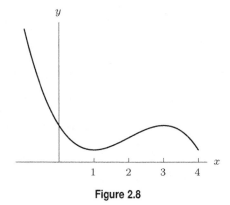

Figure 2.8

29. (a) The only graph in which the slope is 1 for all x is Graph (III).
(b) The only graph in which the slope is positive for all x is Graph (III).
(c) Graphs where the slope is 1 at $x = 2$ are Graphs (III) and (IV).
(d) Graphs where the slope is 2 at $x = 1$ are Graphs (II) and (IV).

Solutions for Section 2.3

1. In Leibniz notation the derivative is dD/dt and the units are feet per minute.

5. (a) The statement $f(5) = 18$ means that when 5 milliliters of catalyst are present, the reaction will take 18 minutes. Thus, the units for 5 are ml while the units for 18 are minutes.

   (b) As in part (a), 5 is measured in ml. Since f' tells how fast T changes per unit a, we have f' measured in minutes/ml. If the amount of catalyst increases by 1 ml (from 5 to 6 ml), the reaction time decreases by about 3 minutes.

9. (a) The units of compliance are units of volume per units of pressure, or liters per centimeter of water.

   (b) The increase in volume for a 5 cm reduction in pressure is largest between 10 and 15 cm. Thus, the compliance appears maximum between 10 and 15 cm of pressure reduction. The derivative is given by the slope, so

$$\text{Compliance} \approx \frac{0.70 - 0.49}{15 - 10} = 0.042 \text{ liters per centimeter.}$$

   (c) When the lung is nearly full, it cannot expand much more to accommodate more air.

13. (a) The units of lapse rate are the same as for the derivative dT/dz, namely (units of T)/(units of z) $=$ $^\circ C/\text{km}$.

   (b) Since the lapse rate is 6.5, the derivative of T with respect to z is $dT/dz = -6.5^\circ C/\text{km}$. The air temperature drops about 6.5° for one more kilometer you go up.

17. (a) The statement $f(200) = 1300$ means that it costs $1300 to produce 200 gallons of the chemical.

   (b) The statement $f'(200) = 6$ means that when the number of gallons produced is 200, costs are increasing at a rate of $6 per gallon. In other words, it costs about $6 to produce the next (the 201st) gallon of the chemical.

21. (a) Since $f'(c)$ is negative, the function $P = f(c)$ is decreasing: on average, pelican eggshells are thinner if the PCB concentration, c, in the eggshell is higher.

   (b) The statement $f(200) = 0.28$ means that the thickness of pelican eggshells is 0.28 mm when the concentration of PCBs in the eggshell is 200 parts per million (ppm).

      The statement $f'(200) = -0.0005$ means that when the PCB concentration is 200 ppm, a 1 ppm increase in the concentration typically corresponds to about a 0.0005 mm decrease in eggshell thickness.

25. (a) The tangent line to the weight graph is steeper at 36 weeks then at 20 weeks, so $g'(36)$ is greater than $g'(20)$.

   (b) The fetus increases its weight more rapidly at week 36 than at week 20.

29. (a) The statement $f(140) = 120$ means that a patient weighing 140 pounds should receive a dose of 120 mg of the painkiller. The statement $f'(140) = 3$ tells us that if the weight of a patient increases by one pound (from 140 pounds), the dose should be increased by about 3 mg.

   (b) Since the dose for a weight of 140 lbs is 120 mg and at this weight the dose goes up by about 3 mg for one pound, a 145 lb patient should get about an additional $3(5) = 15$ mg. Thus, for a 145 lb patient, the correct dose is approximately

$$f(145) \approx 120 + 3(5) = 135 \text{ mg.}$$

33. (a) The slope of the tangent line at 2 kg can be approximated by the slope of the secant line passing through the points $(2, 6)$ and $(3, 2)$. So

$$\text{Slope of tangent line} \approx \frac{v(3) - v(2)}{3 - 2} = \frac{2 - 6}{1} = -4 \text{ (cm/sec)/kg.}$$

   (b) Since 50 grams $= 0.050$ kg, the contraction velocity changes by about $-4(\text{cm/sec})/\text{kg} \cdot 0.050\text{kg} = -0.20$ cm/sec. The velocity is reduced by about 0.20 cm/sec or 2.0 mm/sec.

   (c) Since $v(x)$ is the contraction velocity in cm/sec with a load of x kg, we have $v'(2) = -4$.

37. Where the graph is linear, the derivative of the fat storage function is constant. The derivative gives the rate of fat consumption (kg/week). Thus, for the first four weeks the body burns fat at a constant rate.

41. (a) See part (b).

   (b) See Figure 2.9.

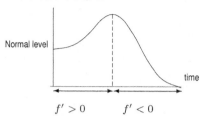

concentration of enzymes

Normal level

time

$f' > 0$ $f' < 0$

Figure 2.9

(c) The derivative, f', is the rate at which the concentration is increasing or decreasing. We see that f' is positive at the start of the disease and negative toward the end. In practice, of course, f' cannot be measured directly. Checking the value of C in blood samples taken on consecutive days allows us to estimate $f'(t)$:

$$f'(t) \approx f(t+1) - f(t) = \frac{f(t+1) - f(t)}{(t+1) - t}.$$

45. (a) Let $f(t)$ be the volume, in cubic km, of the Greenland Ice Sheet t years since 2011 (Alternatively, in year t). We are given information about $f'(t)$, which has unit km^3 per year.
   (b) If t is in years since 2011, we know $f'(0)$ is between -224 and -82 cubic km/year. (Alternatively, $f'(2011)$ is between -224 and -82.)

49. Estimating the relative rate of change using $\Delta t = 0.01$, we have

$$\frac{1}{f} \frac{\Delta f}{\Delta t} \approx \frac{1}{f(10)} \frac{f(10.01) - f(10)}{0.01} = \frac{1}{10^2} \frac{10.01^2 - 10^2}{0.01} = 0.20.$$

53. December 2012 corresponds to $t = 3$. Let $B = f(t)$. The relative rate of change of f at $t = 3$ is $f'(3)/f(3)$. We estimate $f'(3)$ using a difference quotient.
   (a) Estimating the relative rate of change using $\Delta t = 1$ at $t = 3$, we have

$$\frac{dB/dt}{B} = \frac{f'(3)}{f(3)} \approx \frac{1}{f(3)} \frac{f(4) - f(3)}{1} = 0.061 = 6.1\% \text{ per month}$$

   (b) With $\Delta t = 0.1$ and $t = 3$, we have

$$\frac{dB/dt}{B} = \frac{f'(3)}{f(3)} \approx \frac{1}{f(3)} \frac{f(3.1) - f(3)}{0.1} = 0.059 = 5.9\% \text{ per month}$$

   (c) With $\Delta t = 0.01$ and $t = 3$, we have

$$\frac{dB/dt}{B} = \frac{f'(3)}{f(3)} \approx \frac{1}{f(3)} \frac{f(3.01) - f(3)}{0.01} = 0.059 = 5.9\% \text{ per month}$$

The relative rate of change is approximately 5.9% per month.

Solutions for Section 2.4

1. (a) Since $g(x)$ is decreasing at $x = 0$, the value of $g'(0)$ is negative.
   (b) Since $g(x)$ is concave down at $x = 0$, the value of $g''(0)$ is negative.
5. $f'(x) = 0$
   $f''(x) = 0$
9. $f'(x) < 0$
   $f''(x) < 0$

13. The derivative, $s'(t)$, appears to be positive since $s(t)$ is increasing over the interval given. The second derivative also appears to be positive or zero since the function is concave up or possibly linear between $t = 1$ and $t = 3$, i.e., it is increasing at a non-decreasing rate.

17. (a) The derivative, $f'(t)$, appears to be positive since the number of cars is increasing. The second derivative, $f''(t)$, appears to be negative during the period 1975–1990 because the rate of change is increasing. For example, between 1975 and 1980, the rate of change is $(121.6 - 106.7)/5 = 2.98$ million cars per year, while between 1985 and 1990, the rate of change is 1.16 million cars per year.

(b) The derivative, $f'(t)$, appears to be negative between 1990 and 1995 since the number of cars is decreasing, but increasing between 1995 and 2000. The second derivative, $f''(t)$, appears to be positive during the period 1990–2000 because the rate of change is increasing. For example, between 1990 and 1995, the rate of change is $(128.4 - 133.7)/5 = -1.06$ million cars per year, while between 1995 and 2000, the rate of change is 1.04 million cars per year.

(c) To estimate $f'(2005)$ we consider the interval 2000–2005

$$f'(2005) \approx \frac{f(2005) - f(2000)}{2005 - 2000} \approx \frac{136.6 - 133.6}{5} = \frac{3}{5} = 0.6.$$

We estimate that $f'(2005) \approx 0.6$ million cars per year. The number of passenger cars in the US was increasing at a rate of about 600,000 cars per year in 2005.

21. (b). The positive first derivative tells us that the temperature is increasing; the negative second derivative tells us that the rate of increase of the temperature is slowing.

25. (a) Let $N(t)$ be the number of people below the poverty line. See Figure 2.10.

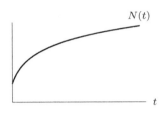

Figure 2.10

(b) dN/dt is positive, since people are still slipping below the poverty line. d^2N/dt^2 is negative, since the rate at which people are slipping below the poverty line, dN/dt, is decreasing.

29. (a) Since the sea level is rising, we know that $a'(t) > 0$ and $m'(t) > 0$. Since the rate is accelerating, we know that $a''(t) > 0$ and $m''(t) > 0$.

(b) The rate of change of sea level for the mid-Atlantic states is between 2 and 4, we know $2 < a'(t) < 4$. (Possibly also $a'(t) = 2$ or $a'(t) = 4$.)
Similarly, $2 < m'(t) < 10$. (Possibly also $m'(t) = 2$ or $m'(t) = 10$.)

(c)   (i) If $a'(t) = 2$, then sea level rise $= 2 \cdot 100 = 200$ mm.
If $a'(t) = 4$, then sea level rise $= 4 \cdot 100 = 400$ mm.
So sea level rise is between 200 mm and 400 mm.

(ii) The shortest amount of time for the sea level in the Gulf of Mexico to rise 1 meter occurs when the rate is largest, 10 mm per year. Since 1 meter = 1000 mm,
shortest time to rise 1 meter $= 1000/10 = 100$ years.

Solutions for Section 2.5

1. The marginal cost is approximated by the difference quotient

$$MC \approx \frac{\Delta C}{\Delta q} = \frac{4830 - 4800}{1305 - 1295} = 3.$$

The marginal cost is approximately $3 per item.

5. Drawing in the tangent line at the point $(10000, C(10000))$ we get Figure 2.11.

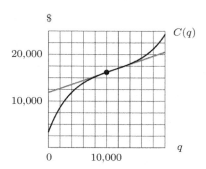

Figure 2.11

We see that each vertical increase of 2500 in the tangent line gives a corresponding horizontal increase of roughly 6000. Thus the marginal cost at the production level of 10,000 units is

$$C'(10{,}000) = \frac{\text{Slope of tangent line}}{\text{to } C(q) \text{ at } q = 10{,}000} = \frac{2500}{6000} = 0.42.$$

This tells us that after producing 10,000 units, it will cost roughly $0.42 to produce one more unit.

9. (a) We can approximate $C(16)$ by adding $C'(15)$ to $C(15)$, since $C'(15)$ is an estimate of the cost of the 16th item.

$$C(16) \approx C(15) + C'(15) = \$2300 + \$108 = \$2408.$$

(b) We approximate $C(14)$ by subtracting $C'(15)$ from $C(15)$, where $C'(15)$ is an approximation of the cost of producing the 15th item.

$$C(14) \approx C(15) - C'(15) = \$2300 - \$108 = \$2192.$$

13. (a) At $q = 2.1$ million,

$$\text{Profit } = \pi(2.1) = R(2.1) - C(2.1) = 6.9 - 5.1 = 1.8 \text{ million dollars.}$$

(b) If $\Delta q = 0.04$,

$$\text{Change in revenue, } \Delta R \approx R'(2.1)\Delta q = 0.7(0.04) = 0.028 \text{ million dollars } = \$28{,}000.$$

Thus, revenues increase by about $28,000.

(c) If $\Delta q = -0.05$,

$$\text{Change in revenue, } \Delta R \approx R'(2.1)\Delta q = 0.7(-0.05) = -0.035 \text{ million dollars } = -\$35{,}000.$$

Thus, revenues decrease by about $35,000.

(d) We find the change in cost by a similar calculation. For $\Delta q = 0.04$,

$$\text{Change in cost, } \Delta C \approx C'(2.1)\Delta q = 0.6(0.04) = 0.024 \text{ million dollars } = \$24{,}000$$
$$\text{Change in profit, } \Delta\pi \approx \$28{,}000 - \$24{,}000 = \$4000.$$

Thus, increasing production 0.04 million units increases profits by about $4000.
For $\Delta q = -0.05$,

$$\text{Change in cost, } \Delta C \approx C'(2.1)\Delta q = 0.6(-0.05) = -0.03 \text{ million dollars } = -\$30{,}000$$
$$\text{Change in profit, } \Delta\pi \approx -\$35{,}000 - (-\$30{,}000) = -\$5000.$$

Thus, decreasing production 0.05 million units decreases profits by about $5000.

Solutions for Chapter 2 Review

1. (a) Let $s = f(t)$.
    (i) We wish to find the average velocity between $t = 1$ and $t = 1.1$. We have

    $$\text{Average velocity} = \frac{f(1.1) - f(1)}{1.1 - 1} = \frac{3.63 - 3}{0.1} = 6.3 \text{ m/sec.}$$

    (ii) We have

    $$\text{Average velocity} = \frac{f(1.01) - f(1)}{1.01 - 1} = \frac{3.0603 - 3}{0.01} = 6.03 \text{ m/sec.}$$

    (iii) We have

    $$\text{Average velocity} = \frac{f(1.001) - f(1)}{1.001 - 1} = \frac{3.006003 - 3}{0.001} = 6.003 \text{ m/sec.}$$

    (b) We see in part (a) that as we choose a smaller and smaller interval around $t = 1$ the average velocity appears to be getting closer and closer to 6, so we estimate the instantaneous velocity at $t = 1$ to be 6 m/sec.

5. (a) $f'(x)$ is negative when the function is decreasing and positive when the function is increasing. Therefore, $f'(x)$ is positive at C and G. $f'(x)$ is negative at A and E. $f'(x)$ is zero at B, D, and F.
    (b) $f'(x)$ is the largest when the graph of the function is increasing the fastest (i.e. the point with the steepest positive slope). This occurs at point G. $f'(x)$ is the most negative when the graph of the function is decreasing the fastest (i.e. the point with the steepest negative slope). This occurs at point A.

9. We use the interval $x = 2$ to $x = 2.01$:

$$f'(2) \approx \frac{f(2.01) - f(2)}{2.01 - 2} = \frac{5^{2.01} - 5^2}{0.01} = \frac{25.4056 - 25}{0.01} = 40.56.$$

For greater accuracy, we can use the smaller interval $x = 2$ to $x = 2.001$:

$$f'(2) \approx \frac{f(2.001) - f(2)}{2.001 - 2} = \frac{5^{2.001} - 5^2}{0.001} = \frac{25.040268 - 25}{0.001} = 40.268.$$

13. See Figure 2.12.

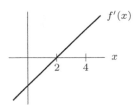

Figure 2.12

17. See Figure 2.13.

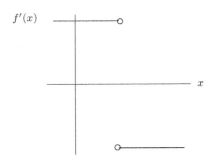

Figure 2.13

21. **(a)** We use the interval to the right of $x = 2$ to estimate the derivative. (Alternately, we could use the interval to the left of 2, or we could use both and average the results.) We have

$$f'(2) \approx \frac{f(4) - f(2)}{4 - 2} = \frac{24 - 18}{4 - 2} = \frac{6}{2} = 3.$$

We estimate $f'(2) \approx 3$.

(b) We know that $f'(x)$ is positive when $f(x)$ is increasing and negative when $f(x)$ is decreasing, so it appears that $f'(x)$ is positive for $0 < x < 4$ and is negative for $4 < x < 12$.

25. (Note that we are considering the average temperature of the yam, since its temperature is different at different points inside it.)

(a) It is positive, because the temperature of the yam increases the longer it sits in the oven.

(b) The units of $f'(20)$ are °F/min. The statement $f'(20) = 2$ means that at time $t = 20$ minutes, the temperature T would increase by approximately $2°$F if the yam is in the oven an additional minute.

29. Moving away slightly from the center of the hurricane from a point 15 kilometers from the center moves you to a point with stronger winds. For example, the wind is stronger at 15.1 kilometers from the center of the hurricane than it is at 15 kilometers from the center.

33. **(a)** If $f'(t) > 0$, the depth of the water is increasing. If $f'(t) < 0$, the depth of the water is decreasing.

(b) The depth of the water is increasing at 20 cm/min when $t = 30$ minutes.

(c) We use 1 meter = 100 cm, 1 hour = 60 min. At time $t = 30$ minutes

$$\text{Rate of change of depth} = 20\frac{\text{cm}}{\text{min}} = 20\frac{\text{cm}}{\text{min}} \cdot \frac{60 \text{ min}}{1 \text{ hr}} \cdot \frac{1 \text{ m}}{100 \text{ cm}} = 12 \text{ meters/hour.}$$

37. The fact that $f(80) = 0.05$ means that when the car is moving at 80 km/hr is it using 0.05 liter of gasoline for each kilometer traveled.

The derivative $f'(v)$ is the rate of change of gasoline consumption with respect to speed. That is, $f'(v)$ tells us how the consumption of gasoline changes as speeds vary. We are told that $f'(80) = 0.0005$. This means that a 1-kilometer increase in speed results in an increase in consumption of about 0.0005 liter per km. At higher speeds, the vehicle burns more gasoline per km traveled than at lower speeds.

41. Units of $P'(t)$ are dollars/year. The practical meaning of $P'(t)$ is the rate at which the monthly payments change as the duration of the mortgage increases. Approximately, $P'(t)$ represents the change in the monthly payment if the duration is increased by one year. $P'(t)$ is negative because increasing the duration of a mortgage decreases the monthly payments.

45. **(a)** minutes/kilometer.

(b) minutes/kilometer^2.

49. **(a)** At t_3, t_4, and t_5, because the graph is above the t-axis there.

(b) At t_2 and t_3, because the graph is sloping up there.

(c) At t_1, t_2, and t_5, because the graph is concave up there

(d) At t_1, t_4, and t_5, because the graph is sloping down there.

(e) At t_3 and t_4, because the graph is concave down there.

STRENGTHEN YOUR UNDERSTANDING

1. True, this is the definition of the derivative.

5. False, the function $f(x) = x^2$ is such a function, with $a = 0$.

9. False, $R(w)$ is increasing for all w so the derivative can not be negative at any point.

13. False. a function can be increasing while its slope is decreasing.

17. False, the function $\ln(t)$ is an increasing function, and therefore has positive derivative.

21. True, since the derivative is the limit of the difference quotient $\frac{f(x+h)-f(x)}{h}$, the units of the derivative are the units of the numerator over the units of the denominator, which is dollars per student.

25. True, dA/dB represents the change in A with respect to a change in B, and so the units are the units of A divided by the units of B.

29. True, since for the first 10 years, height is an increasing function of age.

33. False, since $f'' > 0$ means only that the derivative is increasing, that is, the graph of f is concave up. However, f can still be decreasing, for example $f(x) = 1/x$ for $x > 0$.

37. True, any decaying exponential function has these properties, for example $f(x) = e^{-x}$.

41. True.

45. False, if the revenue function is linear, then the marginal revenue is constant slope of the line, or 5.

49. False, the cost and revenue functions can be equal at q^* but have different slopes.

53. True, since 3% of 100 is 3 and the quantity is decreasing.

Solutions to Problems on Limits and the Definition of the Derivative

1. The answers to parts (a)–(f) are marked in Figure 2.14.

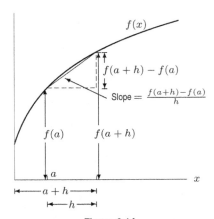

Figure 2.14

5. (a) When we substitute $h = 0$, we get $0/0$. The expression is undefined at $h = 0$.
   (b) We have

$$\text{Substituting } h = 0.1: \quad \frac{\ln(1.1)}{0.1} = 0.953.$$

$$\text{Substituting } h = 0.01: \quad \frac{\ln(1.01)}{0.01} = 0.995.$$

$$\text{Substituting } h = 0.001: \quad \frac{\ln(1.001)}{0.001} = 0.9995.$$

$$\text{Substituting } h = 0.0001: \quad \frac{\ln(1.0001)}{0.0001} = 0.99995.$$

   (c) It appears that

$$\lim_{h \to 0} \frac{\ln(h+1)}{h} \approx 1.$$

9. Using $h = 0.1, 0.01, 0.001$, we see

$$\frac{7^{0.1} - 1}{0.1} = 2.148$$

$$\frac{7^{0.01} - 1}{0.01} = 1.965$$

$$\frac{7^{0.001} - 1}{0.001} = 1.948$$

$$\frac{7^{0.0001} - 1}{0.0001} = 1.946.$$

This suggests that $\lim_{h \to 0} \dfrac{7^h - 1}{h} \approx 1.9$.

13. No, $f(x)$ is not continuous on $0 \leq x \leq 2$, but it is continuous on the interval $0 \leq x \leq 0.5$.

17. (a) Yes, $f(x)$ is continuous on $1 \leq x \leq 3$.

   (b) Yes, $f(x)$ is continuous on $0.5 \leq x \leq 1.5$.

   (c) No, $f(x)$ is not continuous on $3 \leq x \leq 5$ because of the jump at $x = 4$.

   (d) No, $f(x)$ is not continuous on $2 \leq x \leq 6$ because of the jump at $x = 4$.

21. We have

$$\lim_{h \to 0} \frac{(h+1)^2 - 1}{h} = \lim_{h \to 0} \frac{(h^2 + 2h + 1) - 1}{h} = \lim_{h \to 0} \frac{h^2 + 2h}{h} = \lim_{h \to 0} (h + 2) = 0 + 2 = 2.$$

25. Yes: $f(x) = x^2 + 2$ is a continuous function for all values of x.

29. This function is not continuous. Each time someone is born or dies, the number jumps by one.

33. The time is not a continuous function of position as distance from your starting point, because every time you cross from one time zone into the next, the time jumps by 1 hour.

37. Using the definition of the derivative, we have

$$\begin{aligned}
f'(x) &= \lim_{h \to 0} \frac{f(x+h) - f(x)}{h} = \lim_{h \to 0} \frac{3(x+h)^2 - 3x^2}{h} \\
&= \lim_{h \to 0} \frac{3(x^2 + 2xh + h^2) - 3x^2}{h} \\
&= \lim_{h \to 0} \frac{3x^2 + 6xh + 3h^2 - 3x^2}{h} \\
&= \lim_{h \to 0} \frac{6xh + 3h^2}{h} = \lim_{h \to 0} \frac{h(6x + 3h)}{h}.
\end{aligned}$$

As h gets very close to zero (but not equal to zero), we can cancel the h in the numerator and denominator to leave the following:

$$f'(x) = \lim_{h \to 0} (6x + 3h).$$

As $h \to 0$, we have $f'(x) = 6x$.

41. Using the definition of the derivative, we have

$$\begin{aligned}
f'(x) &= \lim_{h \to 0} \frac{f(x+h) - f(x)}{h} \\
&= \lim_{h \to 0} \frac{-2(x+h)^3 - (-2x^3)}{h} \\
&= \lim_{h \to 0} \frac{-2(x^3 + 3x^2 h + 3xh^2 + h^3) + 2x^3}{h} \\
&= \lim_{h \to 0} \frac{-2x^3 - 6x^2 h - 6xh^2 - 2h^3 + 2x^3}{h} \\
&= \lim_{h \to 0} \frac{-6x^2 h - 6xh^2 - 2h^3}{h}.
\end{aligned}$$

As long as we let h get close to zero without actually equaling zero, we can cancel the h in the numerator and denominator, and we are left with $-6x^2 - 6xh - 2h^2$. Taking the limit as h goes to zero, we get $f'(x) = -6x^2$ since the other two terms go to zero.

CHAPTER THREE

Solutions for Section 3.1

1. $\dfrac{dy}{dx} = 3$

5. $y' = 24t^2$

9. $f'(q) = 3q^2$

13. $\dfrac{dy}{dt} = 24t^2 - 8t + 12.$

17. Since $f(z) = -\dfrac{1}{z^{6.1}} = -z^{-6.1}$, we have $f'(z) = -(-6.1)z^{-7.1} = 6.1z^{-7.1}$.

21. Since $f(x) = \sqrt{\dfrac{1}{x^3}} = \dfrac{1}{x^{3/2}} = x^{-3/2}$, we have $f'(x) = -\dfrac{3}{2}x^{-5/2}$.

25. $y' = 2z - \frac{1}{2z^2}$.

29. Since $h(\theta) = \theta(\theta^{-1/2} - \theta^{-2}) = \theta\theta^{-1/2} - \theta\theta^{-2} = \theta^{1/2} - \theta^{-1}$, we have $h'(\theta) = \dfrac{1}{2}\theta^{-1/2} + \theta^{-2}$.

33. Since

$$v = at^2 + \frac{b}{t^2} = at^2 + bt^{-2}$$

we have

$$\frac{dv}{dt} = 2at + b(-2)t^{-3} = 2at - 2\frac{b}{t^3}.$$

37. Since $h(x) = \dfrac{ax + b}{c} = \dfrac{a}{c}x + \dfrac{b}{c}$, we have $h'(x) = \dfrac{a}{c}$.

41. **(a)** $f'(t) = 2t - 4$.
   (b) $f'(1) = 2(1) - 4 = -2$
   $f'(2) = 2(2) - 4 = 0$
   (c) We see from part (b) that $f'(2) = 0$. This means that the slope of the line tangent to the curve at $x = 2$ is zero. From Figure 3.1, we see that indeed the tangent line is horizontal at the point $(2, 1)$. The fact that $f'(1) = -2$ means that the slope of the line tangent to the curve at $x = 1$ is -2. If we draw a line tangent to the graph at $x = 1$ (the point $(1, 2)$) we see that it does indeed have a slope of -2.

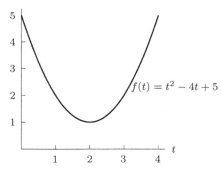

Figure 3.1

45. We have $f'(t) = 6t^2$, so the relative rate of change at $t = 4$ is

$$\frac{f'(4)}{f(4)} = \frac{6(4^2)}{2(4^3) + 10} = \frac{96}{138} = 0.696 = 69.6\% \text{ per year.}$$

49. $f'(t) = 6t^2 - 8t + 3$ and $f''(t) = 12t - 8.$

53. To find the equation of a line we need to have a point on the line and its slope. We know that this line is tangent to the curve $f(t) = 6t - t^2$ at $t = 4$. From this we know that both the curve and the line tangent to it will share the same point and the same slope. At $t = 4$, $f(4) = 6(4) - (4)^2 = 24 - 16 = 8$. Thus we have the point $(4, 8)$. To find the slope, we need to find the derivative. The derivative of $f(t)$ is $f'(t) = 6 - 2t$. The slope of the tangent line at $t = 4$ is $f'(4) = 6 - 2(4) = 6 - 8 = -2$. Now that we have a point and the slope, we can find an equation for the tangent line:

$$y = b + mt$$
$$8 = b + (-2)(4)$$
$$b = 16.$$

Thus, $y = -2t + 16$ is the equation for the line tangent to the curve at $t = 4$. See Figure 3.2.

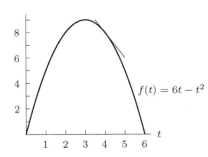

Figure 3.2

57. (a) $A = \pi r^2$
$\frac{dA}{dr} = 2\pi r.$
(b) This is the formula for the circumference of a circle.
(c) $A'(r) \approx \frac{A(r+h) - A(r)}{h}$ for small h. When $h > 0$, the numerator of the difference quotient denotes the area of the region contained between the inner circle (radius r) and the outer circle (radius $r + h$). See figure below. As h approaches 0, this area can be approximated by the product of the circumference of the inner circle and the "width" of the region, i.e., h. Dividing this by the denominator, h, we get $A' =$ the circumference of the circle with radius r.

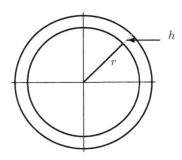

We can also think about the derivative of A as the rate of change of area for a small change in radius. If the radius increases by a tiny amount, the area will increase by a thin ring whose area is simply the circumference at that radius times the small amount. To get the rate of change, we divide by the small amount and obtain the circumference.

61. (a) We have $R(p) = pq = p(300 - 3p) = 300p - 3p^2$
(b) Since $R'(p) = 300 - 6p$, we have $R'(10) = 300 - 6 \cdot 10 = 240$. This means that revenues are increasing at a rate of $240 per dollar of price increase when the price is $10.
(c) $R'(p) = 300 - 6p$ is positive for $p < 50$ and negative for $p > 50$.

65. (a) The marginal cost function equals $C'(q) = 0.08(3q^2) + 75 = 0.24q^2 + 75.$

(b)
$$C(50) = 0.08(50)^3 + 75(50) + 1000 = \$14{,}750.$$

$C(50)$ tells us how much it costs to produce 50 items. From above we can see that the company spends \$14,750 to produce 50 items. The units for $C(q)$ are dollars.

$$C'(50) = 0.24(50)^2 + 75 = \$675 \text{ per item}.$$

$C'(q)$ tells us the approximate change in cost to produce one additional item of product. Thus at $q = 50$ costs will increase by about \$675 for one additional item of product produced. The units are dollars/item.

Solutions for Section 3.2

1. $\dfrac{dP}{dt} = 9t^2 + 2e^t$.

5. $\dfrac{dy}{dx} = 5 \cdot 5^t \ln 5 + 6 \cdot 6^t \ln 6$

9. $\dfrac{dy}{dx} = 5(\ln 2)(2^x) - 5$.

13. $P' = -0.2e^{-0.2t}$.

17. $P'(t) = 12.41(\ln 0.94)(0.94)^t$.

21. Since $y = 10^x + 10x^{-1}$, we have

$$\frac{dy}{dx} = (\ln 10)10^x - 10x^{-2} = (\ln 10)10^x - \frac{10}{x^2}.$$

25. $R'(q) = 2q - 2/q$.

29. Since $f'(t) = 15$, we have

$$\frac{f'(t)}{f(t)} = \frac{15}{15t + 12}.$$

33. Since $f'(t) = -4 \cdot 35t^{-5}$, we have

$$\frac{f'(t)}{f(t)} = \frac{-4 \cdot 35t^{-5}}{35t^{-4}} = -4t^{-1} = -\frac{4}{t}.$$

37. $y = e^{-2t}$, $y' = -2e^{-2t}$. At $t = 0$, $y = 1$ and $y' = -2$. Thus the tangent line at $(0, 1)$ is $y = -2t + 1$.

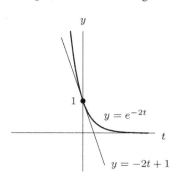

Figure 3.3

41. $f(p) = 10{,}000e^{-0.25p}$, $f(2) = 10{,}000e^{-0.5} = 6065$. If the product sells for \$2, then 6065 units can be sold.

$$f'(p) = 10{,}000e^{-0.25p}(-0.25) = -2500e^{-0.25p}$$

$$f'(2) = -2500e^{-0.5} = -1516.$$

Thus, at a price of \$2, a \$1 increase in price results in a decrease in quantity sold of about 1516 units .

45. Since $P = 35,000(0.98)^t$, the rate of change of the population is given by

$$\frac{dP}{dt} = 35,000 \cdot (\ln 0.98)(0.98^t).$$

On January 1, 2023, we have $t = 13$. At $t = 13$, the rate of change is $35,000(\ln 0.98)(0.98^{13}) = -544$ people/year. The negative sign indicates that the population is decreasing.

49.

$$C(q) = 1000 + 30e^{0.05q}$$
$$C(50) = 1000 + 30e^{2.5} \approx 1365$$

so it costs about \$1365 to produce 50 units.

$$C'(q) = 30(0.05)e^{0.05q} = 1.5e^{0.05q}$$
$$C'(50) = 1.5e^{2.5} \approx 18.27$$

It costs about \$18.27 to produce an additional unit when the production level is 50 units.

53. **(a)** Since the initial population (at $t = 0$) is 1.166 and the growth rate is 1.5%, we have

$$P = 1.166(1 + 0.015)^t = 1.166(1.015)^t \text{ billion.}$$

(b) Differentiating gives

$$\frac{dP}{dt} = 1.166\frac{d}{dt}(1.015)^t = 1.166(1.015)^t(\ln 1.015).$$

$$\left.\frac{dP}{dt}\right|_{t=0} = 1.166(1.015)^0 \ln 1.015 = 0.017 \text{ billion people per year.}$$

$$\left.\frac{dP}{dt}\right|_{t=25} = 1.166(1.015)^{25} \ln 1.015 = 0.025 \text{ billion people per year.}$$

The derivative $\dfrac{dP}{dt}$ is the rate of growth of India's population; $\left.\dfrac{dP}{dt}\right|_{t=0}$ and $\left.\dfrac{dP}{dt}\right|_{t=25}$ are the rates of growth in the years 2009 and 2034, respectively.

Solutions for Section 3.3

1. $f'(x) = 99(x+1)^{98} \cdot 1 = 99(x+1)^{98}$.

5. $\dfrac{dw}{dr} = 3(5r-6)^2 \cdot 5 = 15(5r-6)^2$.

9. $f'(x) = 6(e^{5x})(5) + (e^{-x^2})(-2x) = 30e^{5x} - 2xe^{-x^2}$.

13. $\dfrac{dy}{dt} = \dfrac{5}{5t+1}$.

17. $f'(x) = \dfrac{1}{e^x + 1} \cdot e^x$.

21. $\dfrac{dy}{dx} = \dfrac{1}{3t+2} \cdot 3 = \dfrac{3}{3t+2}$.

25. $\dfrac{dP}{dx} = 0.5(1 + \ln x)^{-0.5}\left(\dfrac{1}{x}\right) = \dfrac{0.5}{x(1+\ln x)^{0.5}}$.

29. We have $f(2) = \ln(2^2 + 1) = \ln(5) = 1.609$. We have $f'(t) = (2t)/(t^2 + 1)$, so $f'(2) = 4/5 = 0.8$. The relative rate of change at $t = 2$ is

$$\frac{f'(2)}{f(2)} = \frac{0.8}{1.609} = 0.497 = 49.7\% \text{ per year.}$$

33. Since $\ln f(t) = \ln 3t^2 = \ln 3 + 2\ln t$ we have

$$\frac{d}{dt}\ln f(t) = \frac{d}{dt}(\ln 3 + 2\ln t) = 0 + \frac{2}{t} = \frac{2}{t}.$$

37. If the distance $s(t) = 20e^{t/2}$, then the velocity, $v(t)$, is given by

$$v(t) = s'(t) = \left(20e^{t/2}\right)' = \left(\frac{1}{2}\right)\left(20e^{t/2}\right) = 10e^{t/2}.$$

41. Estimates may vary. From the graphs, we estimate $g(2) \approx 1.6$, $g'(2) \approx -0.5$, and $f'(1.6) \approx 0.8$. Thus, by the chain rule,

$$h'(2) = f'(g(2)) \cdot g'(2) \approx f'(1.6) \cdot g'(2) \approx 0.8(-0.5) = -0.4.$$

45. The chain rule gives

$$\frac{d}{dx}g(f(x))\bigg|_{x=30} = g'(f(30))f'(30) = g'(20)f'(30) = (1/2)(-2) = -1.$$

49. (a) Differentiating $g(x) = \sqrt{f(x)} = (f(x))^{1/2}$, we have

$$g'(x) = \frac{1}{2}(f(x))^{-1/2} \cdot f'(x) = \frac{f'(x)}{2\sqrt{f(x)}}$$

$$g'(1) = \frac{f'(1)}{2\sqrt{f(1)}} = \frac{3}{2\sqrt{4}} = \frac{3}{4}.$$

(b) Differentiating $h(x) = f(\sqrt{x})$, we have

$$h'(x) = f'(\sqrt{x}) \cdot \frac{1}{2\sqrt{x}}$$

$$h'(1) = f'(\sqrt{1}) \cdot \frac{1}{2\sqrt{1}} = \frac{f'(1)}{2} = \frac{3}{2}.$$

Solutions for Section 3.4

1. By the product rule, $f'(x) = 2x(x^3 + 5) + x^2(3x^2) = 2x^4 + 3x^4 + 10x = 5x^4 + 10x$. Alternatively, $f'(x) = (x^5 + 5x^2)' = 5x^4 + 10x$. The two answers should, and do, match.

5. Differentiating with respect to t, we have

$$\frac{dy}{dt} = \frac{d}{dt}(t^2(3t + 1)^3) = \left(\frac{d}{dt}(t^2)\right)(3t + 1)^3 + t^2\frac{d}{dt}((3t + 1)^3)$$
$$= (2t)(3t + 1)^3 + t^2(3(3t + 1)^2 \cdot 3)$$
$$= 2t(3t + 1)^3 + 9t^2(3t + 1)^2$$

9. $y' = (3t^2 - 14t)e^t + (t^3 - 7t^2 + 1)e^t = (t^3 - 4t^2 - 14t + 1)e^t.$

13. Divide and then differentiate

$$f(x) = x + \frac{3}{x}$$
$$f'(x) = 1 - \frac{3}{x^2}.$$

17. $g'(p) = p\left(\dfrac{2}{2p + 1}\right) + \ln(2p + 1)(1) = \dfrac{2p}{2p + 1} + \ln(2p + 1).$

21. Using the product and chain rules, we have

$$\frac{dz}{dt} = 9(te^{3t} + e^{5t})^8 \cdot \frac{d}{dt}(te^{3t} + e^{5t}) = 9(te^{3t} + e^{5t})^8(1 \cdot e^{3t} + t \cdot e^{3t} \cdot 3 + e^{5t} \cdot 5)$$
$$= 9(te^{3t} + e^{5t})^8(e^{3t} + 3te^{3t} + 5e^{5t}).$$

25. Using the quotient rule,

$$\frac{dw}{dy} = \frac{d}{dy}\left(\frac{3y + y^2}{5 + y}\right) = \frac{(3 + 2y)(5 + y) - (3y + y^2) \cdot 1}{(5 + y)^2}$$
$$= \frac{15 + 13y + 2y^2 - 3y - y^2}{(5 + y)^2} = \frac{15 + 10y + y^2}{(5 + y)^2}.$$

29. We use the product rule. We have $f'(x) = (ax)(e^{-bx}(-b)) + (a)(e^{-bx}) = -abxe^{-bx} + ae^{-bx}$.

33. We multiply out

$$f(x) = (3x + 8)(2x - 5) = 6x^2 + x - 40,$$

and differentiate term by term to get

$$f'(x) = 12x + 1 \quad \text{and} \quad f''(x) = 12.$$

37. (a) See Figure 3.4.

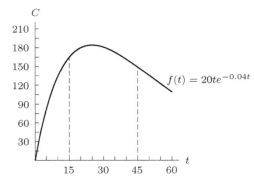

Figure 3.4

Looking at the graph of C, we can see that the see that at $t = 15$, C is increasing. Thus, the slope of the curve at that point is positive, and so $f'(15)$ is also positive. At $t = 45$, the function is decreasing, i.e. the slope of the curve is negative, and thus $f'(45)$ is negative.

(b) We begin by differentiating the function:

$$f'(t) = (20t)(-0.04e^{-0.04t}) + (e^{-0.04t})(20)$$
$$f'(t) = e^{-0.04t}(20 - 0.8t).$$

At $t = 30$,

$$f(30) = 20(30)e^{-0.04 \cdot (30)} = 600e^{-1.2} \approx 181 \text{ mg/ml}$$
$$f'(30) = e^{-1.2}(20 - (0.8)(30)) = e^{-1.2}(-4) \approx -1.2 \text{ mg/ml/min}.$$

These results mean the following: At $t = 30$, or after 30 minutes, the concentration of the drug in the body ($f(30)$) is about 181 mg/ml. The rate of change of the concentration ($f'(30)$) is about -1.2 mg/ml/min, meaning that the concentration of the drug in the body is dropping by 1.2 mg/ml each minute at $t = 30$ minutes.

41. By the product rule, $\frac{d}{dt}tf(t) = f(t) + tf'(t)$. Thus, using the information given in the problem, we have

$$f(t) + tf'(t) = 1 + f(t).$$

Subtracting $f(t)$ from both sides gives $tf'(t) = 1$, so $f'(t) = 1/t$.

45. The chain rule gives

$$(f^n)' = nf^{n-1}f'.$$

Dividing by f^n yields

$$\frac{(f^n)'}{f^n} = \frac{nf^{n-1}f'}{f^n} = n\frac{f'}{f}.$$

Solutions for Section 3.5

1. $\frac{dP}{dt} = -\sin t.$

5. $\frac{dy}{dx} = 5\cos x - 5.$

9. $\frac{dy}{dt} = 2(-\sin(5t))(5) = -10\sin(5t).$

13. $\frac{dy}{dt} = 6 \cdot 2\cos(2t) + (-4\sin(4t)) = 12\cos(2t) - 4\sin(4t).$

17. Using the quotient and chain rules

$$\frac{dz}{dt} = \frac{\frac{d}{dt}(e^{t^2} + t) \cdot \sin(2t) - (e^{t^2} + t)\frac{d}{dt}(\sin(2t))}{(\sin(2t))^2}$$

$$= \frac{\left(e^{t^2} \cdot \frac{d}{dt}(t^2) + 1\right)\sin(2t) - (e^{t^2} + t)\cos(2t)\frac{d}{dt}(2t)}{\sin^2(2t)}$$

$$= \frac{(2te^{t^2} + 1)\sin(2t) - (e^{t^2} + t)2\cos(2t)}{\sin^2(2t)}.$$

21. At $x = \pi$, $y = \sin\pi = 0$, and the slope $\frac{dy}{dx}\Big|_{x=\pi} = \cos x\Big|_{x=\pi} = -1$. Therefore the equation of the tangent line is $y = -(x - \pi) = -x + \pi$. See Figure 3.5.

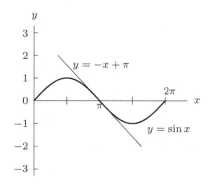

Figure 3.5

25. (a) One cycle is completed in $60/12 = 5$ seconds.

   (b) We differentiate to get

$$A'(t) = 2\left[\sin\left(\frac{2\pi}{5}t\right)\right]\left(\frac{2\pi}{5}\right) = \frac{4\pi}{5}\sin\left(\frac{2\pi}{5}t\right).$$

Substitute $t = 1$ to get

$$A'(1) = \frac{4\pi}{5}\sin\left(\frac{2\pi}{5}(1)\right) = 2.390 \text{ hundred cubic centimeters/second.}$$

This tell us that, one second after the cycle begins, the patient is inhaling at a rate of approximately 2.39 hundred cubic centimeters/second; that is 239 cubic centimeters/second.

29. (a) We substitute $t = 40$:

$$D(t) = 4\cos\left(\frac{2\pi}{365}(t - 172)\right) + 12$$

$$D(40) = 4\cos\left(\frac{2\pi}{365}(40 - 172)\right) + 12 = 9.4186 \text{ hours.}$$

This tells us that, on the 40th day of the year, February 9, 2009, Paris had approximately 9.4 hours of daylight. We differentiate to get

$$D'(t) = -4\left[\sin\left(\frac{2\pi}{365}(t - 172)\right)\right]\left(\frac{2\pi}{365}\right)$$

$$= -\frac{8\pi}{365}\sin\left(\frac{2\pi}{365}(t - 172)\right).$$

Substitute $t = 40$ to get

$$D'(40) = -\frac{8\pi}{365}\sin\left(\frac{2\pi}{365}(40 - 172)\right) = 0.053 \text{ hours/day.}$$

This tell us that, on February 9th, the number of hours of daylight in Paris was increasing at a rate of about 0.053 hours/day. (This is approximately 3.2 minutes per day.)

   (b) We substitute $t = 172$ into $D(t)$ to get

$$D(172) = 4\cos\left(\frac{2\pi}{365}(172 - 172)\right) + 12 = 4\cos(0) + 12 = 16 \text{ hours.}$$

This tells us that, on the 172nd day of the year, June 21, 2009, Paris had approximately 16 hours of daylight. Substitute $t = 172$ into $D'(t)$ to get

$$D'(172) = -\frac{8\pi}{365}\sin\left(\frac{2\pi}{365}(172 - 172)\right) = -\frac{8\pi}{365}\sin(0) = 0 \text{ hours/day.}$$

This tell us that, on June 21st, the rate of change of the number of hours of daylight in Paris was zero. June 21st was the summer solstice (the longest day of the year), so the maximum number of hours of daylight in Paris in 2009 was about 16 hours.

Solutions for Chapter 3 Review

1. $f'(t) = 24t^3$.

5. $\dfrac{dC}{dq} = 0.08e^{0.08q}$.

9. $\dfrac{d}{dt}e^{(1+3t)^2} = e^{(1+3t)^2}\dfrac{d}{dt}(1 + 3t)^2 = e^{(1+3t)^2}\cdot 2(1 + 3t)\cdot 3 = 6(1 + 3t)e^{(1+3t)^2}.$

13. $f'(x) = 6(3(5x - 1)^2) \cdot \dfrac{d}{dx}(5x - 1) = 18(5x - 1)^2(5) = 90(5x - 1)^2$.

17. $\dfrac{dy}{dx} = 2x \ln x + x^2 \cdot \dfrac{1}{x} = x(2 \ln x + 1)$.

21. $h'(t) = \dfrac{1}{e^{-t} - t}\left(-e^{-t} - 1\right)$.

25. $\dfrac{dy}{dx} = 2x \cos x + x^2(-\sin x) = 2x \cos x - x^2 \sin x$.

29. This is a quotient where $u(x) = 1 + e^x$ and $v(x) = 1 - e^{-x}$ so that $q(x) = u(x)/v(x)$.

Using the quotient rule the derivative is

$$q'(x) = \frac{vu' - uv'}{v^2},$$

where $u' = e^x$ and $v' = e^{-x}$. Therefore

$$q'(x) = \frac{(1 - e^{-x})e^x - e^{-x}(1 + e^x)}{(1 - e^{-x})^2} = \frac{e^x - 2 - e^{-x}}{(1 - e^{-x})^2}.$$

33. We use the chain rule with $z = g(x) = x^3$ as the inside function and $f(z) = \cos z$ as the outside function. Since $g'(x) = 3x^2$ and $f'(z) = -\sin z$, we have

$$h'(x) = \frac{d}{dx}\left(\cos(x^3)\right) = -\sin z \cdot (3x^2) = -3x^2 \sin(x^3).$$

37. $\dfrac{dy}{dx} = \dfrac{1}{3}(\ln 3)3^x - \dfrac{33}{2}(x^{-\frac{3}{2}})$.

41. Figure 3.6 shows the graph of $f(x) = x^2 + 1$.

We have $f'(x) = \frac{d}{dx}(x^2 + 1) = 2x$, thus, $f'(0) = 2(0) = 0$. We check this by seeing in that Figure 3.6 the tangent line at $x = 0$ has slope 0.

We have $f'(1) = 2(1) = 2$, $f'(2) = 2(2) = 4$. and $f'(-1) = 2(-1) = -2$. Thus, the slope is positive at $x = 2$ and $x = 1$, and negative at $x = -1$.

Moreover, it is greater at $x = 2$ than at $x = 1$. This agrees with the graph in Figure 3.6.

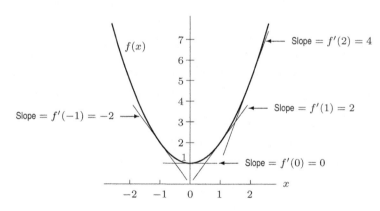

Figure 3.6: Using slopes to check values for derivatives

45. The rate of growth, in billions of people per year, was

$$\frac{dP}{dt} = 6.8(0.012)e^{0.012t}.$$

On January 1, 2009, we have $t = 0$, so

$$\frac{dP}{dt} = 6.8(0.012)e^0 = 0.0816 \text{ billion/year } = 81.6 \text{ million people/year}.$$

49. To find the equation of the line tangent to the graph of $P(t) = t \ln t$ at $t = 2$ we must find the point $(2, P(2))$ as well as the slope of the tangent line at $t = 2$. $P(2) = 2(\ln 2) \approx 1.386$. Thus we have the point $(2, 1.386)$. To find the slope, we must first find $P'(t)$:

$$P'(t) = t\frac{1}{t} + \ln t(1) = 1 + \ln t.$$

At $t = 2$ we have

$$P'(2) = 1 + \ln 2 \approx 1.693$$

Since we now have the slope of the line and a point, we can solve for the equation of the line:

$$Q(t) - 1.386 = 1.693(t - 2)$$
$$Q(t) - 1.386 = 1.693t - 3.386$$
$$Q(t) = 1.693t - 2.$$

The equation of the tangent line is $Q(t) = 1.693t - 2$. We see our results displayed graphically in Figure 3.7.

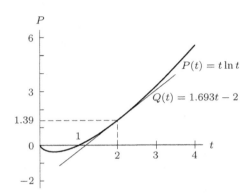

Figure 3.7

53. (a) Differentiating using the chain rule gives

$$\frac{dQ}{dt} = \frac{d}{dt}e^{-0.000121t} = -0.000121e^{-0.000121t}.$$

(b) The following graph shows the rate, dQ/dt, as a function of time.

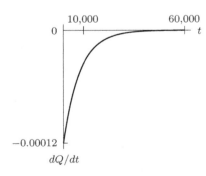

57. (a) To find the temperature of the yam when it was placed in the oven, we need to evaluate the function at $t = 0$. In this case, the temperature of the yam to begin with equals $350(1 - 0.7e^0) = 350(0.3) = 105°$.
   (b) By looking at the function we see that the temperature which the yam is approaching is $350°$. That is, if the yam were left in the oven for a long period of time (i.e. as $t \to \infty$) the temperature would move closer and closer to $350°$ (because $e^{-0.008t}$ would approach zero, and thus $1 - 0.7e^{-0.008t}$ would approach 1). Thus, the temperature of the oven is $350°$.

(c) The yam's temperature will reach $175°$ when $Y(t) = 175$. Thus, we must solve for t:

$$Y(t) = 175$$
$$175 = 350(1 - 0.7e^{-0.008t})$$
$$\frac{175}{350} = 1 - 0.7e^{-0.008t}$$
$$0.7e^{-0.008t} = 0.5$$
$$e^{-0.008t} = 5/7$$
$$\ln e^{-0.008t} = \ln 5/7$$
$$-0.008t = \ln 5/7$$
$$t = \frac{\ln 5/7}{-0.008} \approx 42 \text{ minutes.}$$

Thus the yam's temperature will be $175°$ approximately 42 minutes after it is put into the oven.

(d) The rate at which the temperature is increasing is given by the derivative of the function.

$$Y(t) = 350(1 - 0.7e^{-0.008t}) = 350 - 245e^{-0.008t}.$$

Therefore,

$$Y'(t) = 0 - 245(-0.008e^{-0.008t}) = 1.96e^{-0.008t}.$$

At $t = 20$, the rate of change of the temperature of the yam is given by $Y'(20)$:

$$Y'(20) = 1.96e^{-0.008(20)} = 1.96e^{-.16} = 1.96(0.8521) \approx 1.67 \text{ degrees/minute.}$$

Thus, at $t = 20$ the yam's temperature is increasing by about 1.67 degrees each minute.

61. (a) $H'(2) = r'(2) + s'(2) = -1 + 3 = 2.$
(b) $H'(2) = 5s'(2) = 5(3) = 15.$
(c) $H'(2) = r'(2)s(2) + r(2)s'(2) = -1 \cdot 1 + 4 \cdot 3 = 11.$
(d) $H'(2) = \dfrac{r'(2)}{2\sqrt{r(2)}} = \dfrac{-1}{2\sqrt{4}} = -\dfrac{1}{4}.$

65. Estimates may vary. From the graphs, we estimate $f(1) \approx -0.4$, $f'(1) \approx 0.5$, $g(1) \approx 2$, and $g'(1) \approx 1$. By the product rule,

$$h'(1) = f'(1) \cdot g(1) + f(1) \cdot g'(1) \approx (0.5)2 + (-0.4)1 = 0.6.$$

69. Estimates may vary. From the graphs, we estimate $f(1) \approx -0.4$, $f'(1) \approx 0.5$, $g(1) \approx 2$, and $g'(1) \approx 1$. By the quotient rule, to one decimal place

$$l'(1) = \frac{g'(1) \cdot f(1) - g(1) \cdot f'(1)}{(f(1))^2} \approx \frac{1(-0.4) - 2(0.5)}{(-0.4)^2} = -8.8.$$

73. The first and second derivatives of e^x are e^x. Thus, the graph of $y = e^x$ is concave up. The tangent line at $x = 0$ has slope $e^0 = 1$ and equation $y = x + 1$. A graph that is always concave up is always above any of its tangent lines. Thus $e^x \geq x + 1$ for all x, as shown in Figure 3.8.

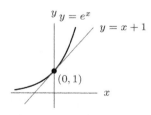

Figure 3.8

77. (a) On the interval $0 < M < 70$, we have

$$\text{Slope} = \frac{\Delta G}{\Delta M} = \frac{2.8}{70} = 0.04 \text{ gallons per mile.}$$

On the interval $70 < M < 100$, we have

$$\text{Slope} = \frac{\Delta G}{\Delta M} = \frac{4.6 - 2.8}{100 - 70} = \frac{1.8}{30} = 0.06 \text{ gallons per mile.}$$

(b) Gas consumption, in miles per gallon, is the reciprocal of the slope, in gallons per mile. On the interval $0 < M < 70$, gas consumption is $1/(0.04) = 25$ miles per gallon. On the interval $70 < M < 100$, gas consumption is $1/(0.06) = 16.667$ miles per gallon.

(c) In Figure 3.26 in the text, we see that the velocity for the first hour of this trip is 70 mph and the velocity for the second hour is 30 mph. The first hour may have been spent driving on an interstate highway and the second hour may have been spent driving in a city. The answers to part (b) would then tell us that this car gets 25 miles to the gallon on the highway and about 17 miles to the gallon in the city.

(d) Since $M = h(t)$, we have $G = f(M) = f(h(t)) = k(t)$. The function k gives the total number of gallons of gas used t hours into the trip. We have

$$G = k(0.5) = f(h(0.5)) = f(35) = 1.4 \text{ gallons.}$$

The car consumes 1.4 gallons of gas during the first half hour of the trip.

(e) Since $k(t) = f(h(t))$, by the chain rule, we have

$$\frac{dG}{dt} = k'(t) = f'(h(t)) \cdot h'(t).$$

Therefore:

$$\left. \frac{dG}{dt} \right|_{t=0.5} = k'(0.5) = f'(h(0.5)) \cdot h'(0.5) = f'(35) \cdot 70 = 0.04 \cdot 70 = 2.8 \text{ gallons per hour,}$$

and

$$\left. \frac{dG}{dt} \right|_{t=1.5} = k'(1.5) = f'(h(1.5)) \cdot h'(1.5) = f'(85) \cdot 30 = 0.06 \cdot 30 = 1.8 \text{ gallons per hour.}$$

Gas is being consumed at a rate of 2.8 gallons per hour at time $t = 0.5$ and is being consumed at a rate of 1.8 gallons per hour at time $t = 1.5$. Notice that gas is being consumed more quickly on the highway, even though the gas mileage is significantly better there.

STRENGTHEN YOUR UNDERSTANDING

1. True. If $f(x) = 5x^2 + 1$ then $f'(x) = 10x$ so $f'(-1) = 10(-1) = -10$.

5. False. The slope of the tangent line is given by the derivative. We have $f'(x) = 5x^4$ so $f'(1) = 5$. The slope of the tangent line at $x = 1$ is 5 (not 9) so the statement is false.

9. True. We see that $f'(x) = 9x^2 - 2x + 2$ and $f''(x) = 18x - 2$. Then $f''(1) = 16$. Since the second derivative is positive at $x = 1$, the graph of f is concave up at $x = 1$.

13. True. We have $f'(x) = 3e^x + 1$ so the slope of the tangent line at $x = 0$ is $f'(0) = 3e^0 + 1 = 3 + 1 = 4$. Since $f(0) = 3e^0 + 0 = 3$, the vertical intercept is 3.

17. False. We see that $f'(x) = 2e^{2x}$.

21. False. The derivative is $f'(t) = 2te^{t^2}$.

25. True.

29. True. The derivative is $f'(x) = -e^{1-x}$ so $f''(x) = e^{1-x}$ and $f''(1) = e^0 = 1$. Since the second derivative is positive at $x = 1$, the function is concave up at $x = 1$.

33. True. The derivative is found using the quotient rule.

37. True. We use the chain rule and then the product rule to find the derivative of the exponent. Notice that $e^{x \ln x} = x^x$ so this method shows us how to find the derivative of x^x.

41. True.

45. False. We use the chain rule to get $y' = -2t \sin t^2$.

49. True.

Solutions to Problems on Establishing the Derivative Formulas ▬▬▬▬▬▬▬

1. Using the definition of the derivative, we have

$$f'(x) = \lim_{h \to 0} \frac{f(x+h) - f(x)}{h}$$

$$= \lim_{h \to 0} \frac{2(x+h) + 1 - (2x+1)}{h}$$

$$= \lim_{h \to 0} \frac{2x + 2h + 1 - 2x - 1}{h}$$

$$= \lim_{h \to 0} \frac{2h}{h}.$$

As long as h is very close to, but not actually equal to, zero we can say that $\lim_{h \to 0} \dfrac{2h}{h} = 2$, and thus conclude that $f'(x) = 2$.

5. The definition of the derivative states that

$$f'(x) = \lim_{h \to 0} \frac{f(x+h) - f(x)}{h}.$$

Using this definition, we have

$$f'(x) = \lim_{h \to 0} \frac{4(x+h)^2 + 1 - (4x^2 + 1)}{h}$$

$$= \lim_{h \to 0} \frac{4x^2 + 8xh + 4h^2 + 1 - 4x^2 - 1}{h}$$

$$= \lim_{h \to 0} \frac{8xh + 4h^2}{h}$$

$$= \lim_{h \to 0} \frac{h(8x + 4h)}{h}.$$

As long as h approaches, but does not equal, zero we can cancel h in the numerator and denominator. The derivative now becomes

$$\lim_{h \to 0} (8x + 4h) = 8x.$$

Thus, $f'(x) = 6x$ as we stated above.

9. Since $f(x) = C$ for all x, we have $f(x + h) = C$. Using the definition of the derivative, we have

$$f'(x) = \lim_{h \to 0} \frac{f(x+h) - f(x)}{h}$$

$$= \lim_{h \to 0} \frac{C - C}{h}$$

$$= \lim_{h \to 0} \frac{0}{h}.$$

As h gets very close to zero without actually equaling zero, we have $0/h = 0$, so

$$f'(x) = \lim_{h \to 0} (0) = 0.$$

Solutions to Practice Problems on Differentiation ▬▬▬▬▬▬▬

1. $f'(t) = 2t + 4t^3$

5. $f'(x) = -2x^{-3} + 5\left(\frac{1}{2}x^{-1/2}\right) = \dfrac{-2}{x^3} + \dfrac{5}{2\sqrt{x}}$

9. $D'(p) = 2pe^{p^2} + 10p$

13. $s'(t) = \dfrac{16}{2t+1}$

17. $C'(q) = 3(2q+1)^2 \cdot 2 = 6(2q+1)^2$

21. $y' = 2x \ln(2x+1) + \dfrac{2x^2}{2x+1}$

25. $g'(t) = 15 \cos(5t)$

29. $y' = 17 + 12x^{-1/2}$.

33. Either notice that $f(x) = \dfrac{x^2 + 3x + 2}{x+1}$ can be written as $f(x) = \dfrac{(x+2)(x+1)}{x+1}$ which reduces to $f(x) = x+2$, giving $f'(x) = 1$, or use the quotient rule which gives

$$
\begin{aligned}
f'(x) &= \frac{(x+1)(2x+3) - (x^2+3x+2)}{(x+1)^2} \\
&= \frac{2x^2 + 5x + 3 - x^2 - 3x - 2}{(x+1)^2} \\
&= \frac{x^2 + 2x + 1}{(x+1)^2} \\
&= \frac{(x+1)^2}{(x+1)^2} \\
&= 1.
\end{aligned}
$$

37. $q'(r) = \dfrac{3(5r+2) - 3r(5)}{(5r+2)^2} = \dfrac{15r + 6 - 15r}{(5r+2)^2} = \dfrac{6}{(5r+2)^2}$

41. $h'(w) = 5(w^4 - 2w)^4(4w^3 - 2)$

45. $h'(w) = 6w^{-4} + \dfrac{3}{2}w^{-1/2}$

49. Using the chain rule, $g'(\theta) = (\cos\theta)e^{\sin\theta}$.

53. $h'(r) = \dfrac{d}{dr}\left(\dfrac{r^2}{2r+1}\right) = \dfrac{(2r)(2r+1) - 2r^2}{(2r+1)^2} = \dfrac{2r(r+1)}{(2r+1)^2}$.

57. $f'(x) = \dfrac{3x^2}{9}(3\ln x - 1) + \dfrac{x^3}{9}\left(\dfrac{3}{x}\right) = x^2 \ln x - \dfrac{x^2}{3} + \dfrac{x^2}{3} = x^2 \ln x$

61. Using the quotient rule gives

$$
\begin{aligned}
w'(r) &= \frac{2ar(b + r^3) - 3r^2(ar^2)}{(b + r^3)^2} \\
&= \frac{2abr - ar^4}{(b + r^3)^2}.
\end{aligned}
$$

CHAPTER FOUR

Solutions for Section 4.1

1. We find a critical point by noting where $f'(x) = 0$ or f' is undefined. Since the curve is smooth throughout, f' is always defined, so we look for where $f'(x) = 0$, or equivalently where the tangent line to the graph is horizontal. These points are shown in Figure 4.1.

Figure 4.1

As we can see, there is one critical point. Since it is higher than nearby points, it is a local maximum.

5. **(a)** One possible answer is shown in Figure 4.2.
   (b) One possible answer is shown in Figure 4.3

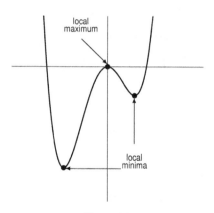

Figure 4.2

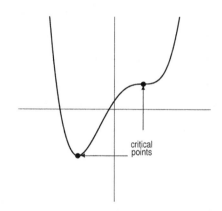

Figure 4.3

9. The graph of f in Figure 4.4 appears to be increasing for $x < -1.4$, decreasing for $-1.4 < x < 1.4$, and increasing for $x > 1.4$. There is a local maximum near $x = -1.4$ and local minimum near $x = 1.4$. The derivative of f is $f'(x) = 3x^2 - 6$. Thus $f'(x) = 0$ when $x^2 = 2$, that is $x = \pm\sqrt{2}$. This explains the critical points near $x = \pm 1.4$. Since $f'(x)$ changes from positive to negative at $x = -\sqrt{2}$, and from negative to positive at $x = \sqrt{2}$, there is a local maximum at $x = -\sqrt{2}$ and a local minimum at $x = \sqrt{2}$.

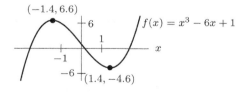

Figure 4.4

13. The graph of f in Figure 4.5 looks like a climbing sine curve, alternately increasing and decreasing, with more time spent increasing than decreasing. Here $f'(x) = 1 + 2\cos x$, so $f'(x) = 0$ when $\cos x = -1/2$; this occurs when

$$x = \pm\frac{2\pi}{3}, \pm\frac{4\pi}{3}, \pm\frac{8\pi}{3}, \pm\frac{10\pi}{3}, \pm\frac{14\pi}{3}, \pm\frac{16\pi}{3}\ldots$$

Since $f'(x)$ changes sign at each of these values, the graph of f must alternate increasing/decreasing. However, the distance between values of x for critical points alternates between $(2\pi)/3$ and $(4\pi)/3$, with $f'(x) > 0$ on the intervals of length $(4\pi)/3$. For example, $f'(x) > 0$ on the interval $(4\pi)/3 < x < (8\pi)/3$. As a result, f is increasing on the intervals of length $(4\pi/3)$ and decreasing on the intervals of length $(2\pi/3)$.

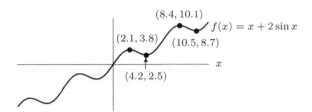

Figure 4.5

17. $f'(x) = 7(x^2 - 4)^6 2x = 14x(x - 2)^6(x + 2)^6$. The critical points of f are $x = 0$, $x = \pm 2$. To the left of $x = -2$, $f'(x) < 0$. Between $x = -2$ and $x = 0$, $f'(x) < 0$. Between $x = 0$ and $x = 2$, $f'(x) > 0$. To the right of $x = 2$, $f'(x) > 0$. Thus, $f(0)$ is a local minimum, whereas $f(-2)$ and $f(2)$ are not local extrema. See Figure 4.6.

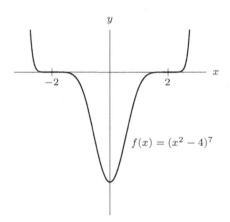

Figure 4.6

21. Differentiating using the product rule gives

$$f'(x) = 3x^2(1 - x)^4 - 4x^3(1 - x)^3 = x^2(1 - x)^3(3(1 - x) - 4x) = x^2(1 - x)^3(3 - 7x).$$

The critical points are the solutions to

$$f'(x) = x^2(1 - x)^3(3 - 7x) = 0$$
$$x = 0, 1, \frac{3}{7}.$$

For $x < 0$, since $1 - x > 0$ and $3 - 7x > 0$, we have $f'(x) > 0$.
For $0 < x < \frac{3}{7}$, since $1 - x > 0$ and $3 - 7x > 0$, we have $f'(x) > 0$.
For $\frac{3}{7} < x < 1$, since $1 - x > 0$ and $3 - 7x < 0$, we have $f'(x) < 0$.
For $1 < x$, since $1 - x < 0$ and $3 - 7x < 0$, we have $f'(x) > 0$.
Thus, $x = 0$ is neither a local maximum nor a local minimum; $x = 3/7$ is a local maximum; $x = 1$ is a local minimum.

25. (a) Increasing for all x.

(b) No maxima or minima.

29. (a) The demand for the product is increasing when $f'(t)$ is positive, and decreasing when $f'(t)$ is negative. Inspection of the table suggests that demand is increasing during weeks 0 to 2 and weeks 6 to 10, and decreasing during weeks 3 to 5.

(b) Because the demand for the product is increasing during weeks 0 to 2, decreasing during weeks 3 to 5, and increasing during weeks 6 to 10, the demand has a local minimum during week 0, a local maximum sometime during week 2 or 3, a local minimum during week 5 or 6, and a local maximum during week 10.

33. If the minimum of $f(x)$ is at $(-2, -3)$, then the derivative of f must be equal to 0 there. In other words, $f'(-2) = 0$. If

$$f(x) = x^2 + ax + b, \quad \text{then}$$
$$f'(x) = 2x + a$$
$$f'(-2) = 2(-2) + a = -4 + a = 0$$

so $a = 4$. Since $(-2, -3)$ is on the graph of $f(x)$ we know that $f(-2) = -3$. So

$$f(-2) = (-2)^2 + a(-2) + b = -3$$
$$a = 4, \text{ so} \quad (-2)^2 + 4(-2) + b = -3$$
$$4 - 8 + b = -3$$
$$-4 + b = -3$$
$$b = 1$$

so $a = 4$ and $b = 1$, and $f(x) = x^2 + 4x + 1$.

37. (a) The function $f(x)$ is defined for $x \geq 0$.

We set the derivative equal to zero and solve for x to find critical points:

$$f'(x) = 1 - \frac{1}{2}ax^{-1/2} = 0$$
$$1 - \frac{a}{2\sqrt{x}} = 0$$
$$2\sqrt{x} = a$$
$$x = \frac{a^2}{4}.$$

Notice that f' is undefined at $x = 0$ so there are two critical points: $x = 0$ and $x = a^2/4$.

(b) We want the critical point $x = a^2/4$ to occur at $x = 5$, so we have:

$$5 = \frac{a^2}{4}$$
$$20 = a^2$$
$$a = \pm\sqrt{20}.$$

Since a is positive, we use the positive square root. The second derivative,

$$f''(x) = \frac{1}{4}ax^{-3/2} = \frac{1}{4}\sqrt{20}x^{-3/2}$$

is positive for all $x > 0$, so the function is concave up and $x = 5$ gives a local minimum. See Figure 4.7.

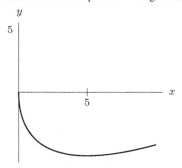

Figure 4.7

41. **(a)** See Figure 4.8.

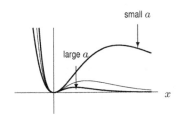

Figure 4.8

(b) We see in Figure 4.8 that in each case f appears to have two critical points. One critical point is a local minimum at the origin and the other is a local maximum in quadrant I. As the parameter a increases, the critical point in quadrant I appears to move down and to the left, closer to the origin.

(c) To find the critical points, we set the derivative equal to zero and solve for x. Using the product rule, we have:

$$f'(x) = x^2 \cdot e^{-ax}(-a) + 2x \cdot e^{-ax} = 0$$
$$xe^{-ax}(-ax + 2) = 0$$
$$x = 0 \quad \text{and} \quad x = \frac{2}{a}.$$

There are two critical points, at $x = 0$ and $x = 2/a$. As we saw in the graph, as a increases the nonzero critical point moves to the left.

Solutions for Section 4.2

1. We find an inflection point by noting where the concavity changes. Such points are shown in Figure 4.9. There are two inflection points.

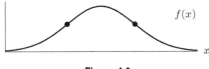

Figure 4.9

5. One possible answer is shown in Figure 4.10.

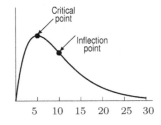

Figure 4.10

9. **(a)** There was a critical point at 6 pm when the temperature was at a local minimum.
(b) The graph of temperature was decreasing but concave up in the morning. In the early afternoon the graph was decreasing but concave down. There was an inflection point at noon when the northerly wind started blowing. By 6 pm when the temperature was at a local minimum, the graph must have been concave up again so there must have been a second inflection point between noon and 6 pm. See Figure 4.11.

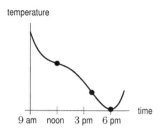

Figure 4.11

13. A critical point will occur whenever $f'(x) = 0$ or f' is undefined. Since $f'(x)$ is always defined, we set

$$f'(x) = 3x^2 - 3 = 3(x^2 - 1) = 0.$$

Factoring, we get

$$f'(x) = 3(x - 1)(x + 1) = 0.$$

So, $x = 1$ or $x = -1$. To find the inflection points of $f(x)$, we find where $f''(x)$ goes from negative to positive or vice versa. For a point to satisfy this condition, it must have at least $f''(x) = 0$ or f'' undefined. Since $f''(x) = 6x$, we know $f''(x)$ is always defined. It is zero when $6x = 0$, so $x = 0$. Since $f''(x) = 6x$ is negative for $x < 0$ and positive for $x > 0$, $x = 0$ must be an inflection point for $f(x)$.

So $x = 1$ and $x = -1$ are critical points of $f(x)$, and $x = 0$ is an inflection point for $f(x)$.

To identify the nature of the critical points $x = 1$ and $x = -1$ that we have found, we can look at a graph of $f(x)$ for values of x near the critical points. Such a graph is shown in Figure 4.12. From the graph we see that $f(-1)$ is a local maximum of f and $f(1)$ is a local minimum of f.

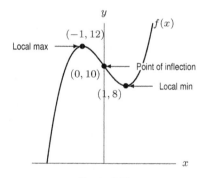

Figure 4.12

17. $f'(x) = 12x^3 - 12x^2$. To find critical points, we set $f'(x) = 0$. This implies $12x^2(x - 1) = 0$. So the critical points of f are $x = 0$ and $x = 1$. To find the inflection points of $f(x)$ we look for points at which $f''(x)$ goes from negative to positive or vice-versa. At any such point $f''(x)$ is either zero or undefined. Since $f''(x) = 36x^2 - 24x = 12x(3x - 2)$, our candidate points are $x = 0$ and $x = 2/3$. Both $x = 0$ and $x = 2/3$ are inflection points, since at both, $f''(x)$ changes sign.

From Figure 4.13, we see that the critical point $x = 1$ is a local minimum and the critical point $x = 0$ is neither a local maximum nor a local minimum.

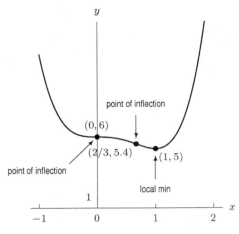

Figure 4.13

21. We see that $f'(x) = 15x^4 - 15x^2$ and $f''(x) = 60x^3 - 30x$. Since $f'(x) = 15x^2(x^2 - 1)$, the critical points are $x = 0, \pm 1$.

   To find possible inflection points, we determine when $f''(x) = 0$. Since $f''(x) = 30x(2x^2 - 1)$, we have $f''(x) = 0$ when $x = 0$ or $x = \pm 1/\sqrt{2}$. Since $f''(x)$ changes sign at each of these points, all are inflection points.

   We see from Figure 4.14 that the critical point $x = -1$ is a local maximum, $x = 1$ is a local minimum, and $x = 0$ is neither.

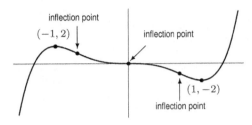

Figure 4.14

25. The inflection points of f are the points where f'' changes sign. See Figure 4.15.

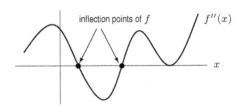

Figure 4.15

29. See Figure 4.16.

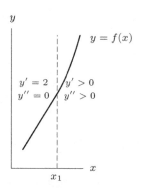

Figure 4.16

33. **(a)** The inflection point occurs at about week 14 where the graph changes from concave up to concave down.
   (b) The fetus increases its length faster at week 14 than at any other time during its gestation.

37. **(a)** The concavity changes at t_1 and t_3, as shown in Figure 4.17.

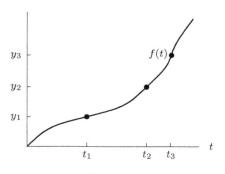

Figure 4.17

   (b) $f(t)$ grows most quickly where the vase is skinniest (at y_3) and most slowly where the vase is widest (at y_1). The diameter of the widest part of the vase looks to be about 4 times as large as the diameter at the skinniest part. Since the area of a cross section is given by πr^2, where r is the radius, the ratio between areas of cross sections at these two places is about 4^2, so the growth rates are in a ratio of about 1 to 16 (the wide part being 16 times slower).

Solutions for Section 4.3

1. See Figure 4.18.

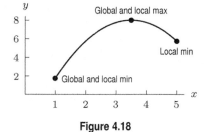

Figure 4.18

5. See Figure 4.19.

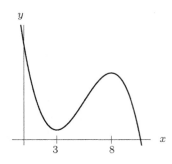

Figure 4.19

9. Using a computer to graph the function, $f(x) = x^3 - e^x$, and its derivative, $f'(x) = 3x^2 - e^x$, we find that the derivative crosses the x-axis three times in the interval $-1 \le x \le 4$ and twice in the interval $-3 \le x \le 2$. See Figures 4.20 and 4.21.

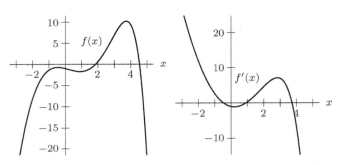

Figure 4.20: Function, f **Figure 4.21:** Derivative, f'

Through trial and error, we obtain approximations: local maximum at $x \approx 3.73$, local minimum at $x \approx 0.91$ and local maximum at $x \approx -0.46$. We can use the approximate values at these points, along with a picture as a guide, to find the global maximum and minimum on any interval.

(a) On the interval $-1 \le x \le 4$, f has local minima at the endpoints $x = -1$ and $x = 4$, in addition to the local extrema at the critical points listed above. We find the global minimum and maximum on the interval by examining the critical points as well as the endpoints. Since $f(-1) = -1.3679$, $f(-0.46) = -0.7286$, $f(0.91) = -1.7308$, $f(3.73) = 10.2160$, $f(4) = 9.4018$, we see $x \approx 0.91$ gives a global minimum on the interval and $x \approx 3.73$ gives a global maximum.

(b) On the interval $-3 \le x \le 2$, f has a local minimum at $x = -3$ and a local maximum at $x = 2$, in addition to the local extrema at the critical points listed above. We find the global minimum and maximum on the interval by examining the critical points as well as the endpoints. Since $f(-3) = -27.0498$, $f(-0.46) = -0.7286$, $f(0.91) = -1.7308$, $f(2) = 0.6109$, we see $x = -3$ gives a global minimum and $x = 2$ a global maximum. (Even though $x \approx -0.46$ gives a local maximum, it does not give the greatest maximum on this interval; even though $x \approx 0.91$ gives a local minimum, it is not the smallest minimum on this interval.)

13. See Figure 4.22.

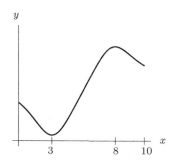

Figure 4.22

17. (a) Differentiating $f(x) = 2x^3 - 9x^2 + 12x + 1$ produces $f'(x) = 6x^2 - 18x + 12$. A second differentiation produces $f''(x) = 12x - 18$.

  (b) $f'(x)$ is defined for all x and $f'(x) = 0$ when $x = 1, 2$. Thus $x = 1, 2$ are critical points.

  (c) $f''(x)$ is defined for all x and $f''(x) = 0$ when $x = \frac{3}{2}$. Since the concavity of f changes at this point, it is an inflection point.

  (d) We have $f(-0.5) = -7.5$, $f(3) = 10$, $f(1) = 6$, $f(2) = 5$. So f has a local maximum at $x = 1$ and at $x = 3$ and a local minimum at $x = -0.5$ and at $x = 2$, a global maximum at $x = 3$ and a global minimum at $x = -0.5$

  (e) Plotting the function $f(x)$ for $-0.5 \leq x \leq 3$ gives the graph shown in Figure 4.23.

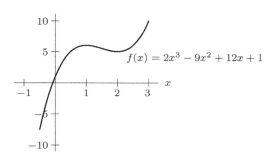

Figure 4.23

21. We find all critical points:

$$\frac{dy}{dx} = 2ax + b = 0$$

$$x = -\frac{b}{2a}.$$

Since y is a quadratic polynomial, its graph is a parabola which opens up if $a > 0$ and down if $a < 0$. The critical value is a maximum if $a < 0$ and a minimum if $a > 0$.

25. Let the numbers be x, y, z and let $y = 2x$. Then

$$x + y + z = 3x + z = 36, \quad \text{so} \quad z = 36 - 3x.$$

Since all the numbers are nonnegative, we restrict to $0 \leq x \leq 12$.

  The product is

$$P = xyz = x \cdot 2x \cdot (36 - 3x) = 72x^2 - 6x^3.$$

Differentiating to find the maximum,

$$\frac{dP}{dx} = 144x - 18x^2 = 0$$

$$-18x(x - 8) = 0$$

$$x = 0, 8.$$

So there are critical points at $x = 0$ and $x = 8$; the end points are at $x = 0, 12$.

Evaluating gives:

At $x = 0$, we have $P = 0$.

At $x = 8$, we have $P = 8 \cdot 16 \cdot (36 - 3 \cdot 8) = 1536$.

At $x = 12$, we have $P = 12 \cdot 24 \cdot (36 - 3 \cdot 12) = 0$.

Thus, the maximum value of the product is 1536.

29. Differentiating using the product rule gives

$$g'(t) = 1 \cdot e^{-t} - te^{-t} = (1 - t)e^{-t},$$

so the critical point is $t = 1$.

Since $g'(t) > 0$ for $0 < t < 1$ and $g'(t) < 0$ for $t > 1$, the critical point is a local maximum.

As $t \to \infty$, the value of $g(t) \to 0$, and as $t \to 0^+$, the value of $g(t) \to 0$. Thus, the local maximum at $x = 1$ is a global maximum of $g(1) = 1e^{-1} = 1/e$. In addition, the value of $g(t)$ is positive for all $t > 0$; it tends to 0 but never reaches 0. Thus, there is no global minimum. See Figure 4.24.

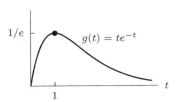

Figure 4.24

33. To find the value of w that minimizes S, we set dS/dw equal to zero and solve for w. To find dS/dw, we first solve for S:

$$S - 5pw = 3qw^2 - 6pq$$
$$S = 5pw + 3qw^2 - 6pq.$$

We now find the critical points:

$$\frac{dS}{dw} = 5p + 6qw = 0$$
$$w = -\frac{5p}{6q}.$$

There is one critical point. Since S is a quadratic function of w with a positive leading coefficient, the function has a minimum at this critical point.

37. **(a)** Suppose the farmer plants x trees per km^2. Then, for $x \leq 200$, the yield per tree is 400 kg, so

$$\text{Total yield, } y = 400x \text{ kg.}$$

For $x > 200$, the yield per tree is reduced by 1 kg for each tree over 200, so the yield per tree is $400 - (x - 200)$ kg $= (600 - x)$ kg. Thus,

$$\text{Total yield, } y = (600 - x)x = 600x - x^2 \text{ kg.}$$

The graph of total yield is in Figure 4.25. It is a straight line for $x \leq 200$ and a parabola for $x > 200$.

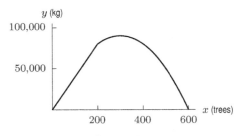

Figure 4.25

(b) The maximum occurs where the derivative of the quadratic is zero, or

$$\frac{dy}{dx} = 600 - 2x = 0$$

$$x = 300 \text{ trees/km}^2.$$

41. (a) The variables are S and H. At the maximum value of S, using the product rule, we have

$$\frac{dS}{dH} = ae^{-bH} + aH(-b)e^{-bH} = 0$$

$$ae^{-bH}(1 - bH) = 0$$

$$bH = 1$$

$$H = \frac{1}{b}.$$

Thus, the maximum value of S occurs when $H = 1/b$. To find the maximum value of S, we substitute $H = 1/b$, giving

$$S = a\frac{1}{b}e^{-b(1/b)} = \frac{a}{b}e^{-1}.$$

(b) Increasing a increases the maximum value of S. Increasing b decreases the maximum value of S.

45. (a) The maximum and minimum values of p can be found without taking derivatives, since the function $20\sin(2.5\pi t)$ has maximum and minimum values of 20 and -20, respectively. Thus, the maximum value of p is 120 mm Hg and the minimum value is 80 mm Hg.

(b) The time between successive maxima is the period, which is $2\pi/(2.5\pi) = 0.8$ seconds.

(c) See Figure 4.26.

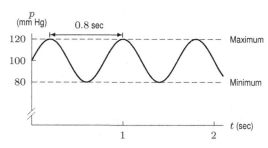

Figure 4.26

Solutions for Section 4.4

1. The profit function is positive when $R(q) > C(q)$, and negative when $C(q) > R(q)$. It's positive for $5.5 < q < 12.5$, and negative for $0 < q < 5.5$ and $12.5 < q$. Profit is maximized when $R(q) > C(q)$ and $R'(q) = C'(q)$ which occurs at about $q = 9.5$. See Figure 4.27.

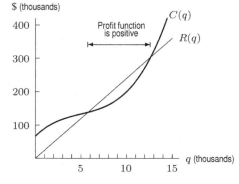

Figure 4.27

5. (a) See Figure 4.28. We find q_1 and q_2 by checking to see where the slope of the tangent line to $C(q)$ is equal to the slope of $R(q)$. Because the slopes of $C(q)$ and $R(q)$ represent marginal cost and marginal revenue, respectively, at q's where the slopes are equal, the cost of producing an additional unit of q exactly equals the revenue gained from selling an additional unit. In other words, q_1 and q_2 are levels of production at which the additional or marginal profit from producing an additional unit of q is zero.

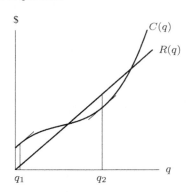

Figure 4.28

(b) We can think of the vertical distance between the cost and revenue curves as representing the firm's total profits. The quantity q_1 is the production level at which total profits are at a minimum. We see this by noticing that for points slightly to the left of q_1, the slope of $C(q)$ is slightly greater than the slope of $R(q)$. This means that the cost of producing an additional unit of q is greater than the revenue earned from selling it. So for points to the left of q_1, additional production decreases profits. For points slightly to the right of q_1, the slope of $C(q)$ is less than the slope of $R(q)$. Thus additional production results in an increase in profits. The production level q_1 is the level at which the firm ceases to take a loss on each additional item and begins to make a profit on each additional item. Note that the total profit is still negative, and remains so until the graphs cross, i.e, where total cost equals total revenue.

Similar reasoning applies for q_2, except it is the level of production at which profits are maximized. For points slightly to the left of q_2, the slope of $C(q)$ is less than the slope of $R(q)$. Thus the cost of producing an additional unit is less than the revenue gained from selling it. So by selling an additional unit, the firm can increase profits. For points to the right of q_2, the slope of $C(q)$ is greater than the slope of $R(q)$. This means that the profit from producing and selling an additional unit of q will be negative, decreasing total profits. The point q_2 is the level of production at which the firm stops making a profit on each additional item sold and begins to take a loss.

At both points q_1 and q_2, note that the vertical distance between $C(q)$ and $R(q)$ is at a local maximum. This represents the fact that q_1 and q_2 are local profit minimum and local profit maximum points.

9. (a) At $q = 5000$, $MR > MC$, so the marginal revenue to produce the next item is greater than the marginal cost. This means that the company will make money by producing additional units, and production should be increased.

(b) Profit is maximized where $MR = MC$, and where the profit function is going from increasing ($MR > MC$) to decreasing ($MR < MC$). This occurs at $q = 8000$.

13. The profit is maximized at the point where the difference between revenue and cost is greatest. Thus the profit is maximized at approximately $q = 4000$.

17. The marginal revenue, MR, is given by differentiating the total revenue function, R. We use the chain rule so

$$MR = \frac{dR}{dq} = \frac{1}{1 + 1000q^2} \cdot \frac{d}{dq}\left(1000q^2\right) = \frac{1}{1 + 1000q^2} \cdot 2000q.$$

When $q = 10$,

$$\text{Marginal revenue} = \frac{2000 \cdot 10}{1 + 1000 \cdot 10^2} = \$0.20/\text{item}.$$

When 10 items are produced, each additional item produced gives approximately $0.20 in additional revenue.

21. We first need to find an expression for revenue in terms of price. At a price of $8, 1500 tickets are sold. For each $1 above $8, 75 fewer tickets are sold. This suggests the following formula for q, the quantity sold for any price p.

$$q = 1500 - 75(p - 8)$$
$$= 1500 - 75p + 600$$
$$= 2100 - 75p.$$

We know that $R = pq$, so substitution yields

$$R(p) = p(2100 - 75p) = 2100p - 75p^2$$

To maximize revenue, we find the derivative of $R(p)$ and set it equal to 0.

$$R'(p) = 2100 - 150p = 0$$
$$150p = 2100$$

so $p = \frac{2100}{150} = 14$. Does $R(p)$ have a maximum at $p = 14$? Using the first derivative test,

$$R'(p) > 0 \ \text{ if } \ p < 14 \text{ and}$$
$$R'(p) < 0 \ \text{ if } \ p > 14.$$

So $R(p)$ has a local maximum at $p = 14$. Since this is the only critical point for $p \geq 0$, it must be a global maximum. So we conclude that revenue is maximized when the price is \$14.

25. Consider the rectangle of sides x and y shown in Figure 4.29.

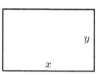

Figure 4.29

The total area is $xy = 3000$, so $y = 3000/x$. Suppose the left and right edges and the lower edge have the shrubs and the top edge has the fencing. The total cost is

$$C = 45(x + 2y) + 20(x)$$
$$= 65x + 90y.$$

Since $y = 3000/x$, this reduces to

$$C(x) = 65x + 90(3000/x) = 65x + 270{,}000/x.$$

Therefore, $C'(x) = 65 - 270{,}000/x^2$. We set this to 0 to find the critical points:

$$65 - \frac{270{,}000}{x^2} = 0$$
$$\frac{270{,}000}{x^2} = 65$$
$$x^2 = 4153.85$$
$$x = 64.450 \text{ ft}$$

so that

$$y = 3000/x = 46.548 \text{ ft.}$$

Since $C(x) \to \infty$ as $x \to 0^+$ and $x \to \infty$, we see that $x = 64.450$ is a minimum. The minimum total cost is then

$$C(64.450) \approx \$8378.54.$$

29. (a) The business must reorder often enough to keep pace with sales. If reordering is done every t months, then,

$$\text{Quantity sold in } t \text{ months} = \text{Quantity reordered in each batch}$$
$$rt = q$$
$$t = \frac{q}{r} \text{ months.}$$

(b) The amount spent on each order is $a + bq$, which is spent every q/r months. To find the monthly expenditures, divide by q/r. Thus, on average,

$$\text{Amount spent on ordering per month} = \frac{a + bq}{q/r} = \frac{ra}{q} + rb \text{ dollars.}$$

(c) The monthly cost of storage is $kq/2$ dollars, so

$$C = \text{Ordering costs} + \text{Storage costs}$$
$$C = \frac{ra}{q} + rb + \frac{kq}{2} \text{ dollars.}$$

(d) The optimal batch size minimizes C, so

$$\frac{dC}{dq} = \frac{-ra}{q^2} + \frac{k}{2} = 0$$
$$\frac{ra}{q^2} = \frac{k}{2}$$
$$q^2 = \frac{2ra}{k}$$

so

$$q = \sqrt{\frac{2ra}{k}} \text{ items per order.}$$

Solutions for Section 4.5

1. (a) Since the graph is concave down, the average cost gets smaller as q increases. This is because the cost per item gets smaller as q increases. There is no value of q for which the average cost is minimized since for any q_0 larger than q the average cost at q_0 is less than the average cost at q. Graphically, the average cost at q is the slope of the line going through the origin and through the point $(q, C(q))$. Figure 4.30 shows how as q gets larger, the average cost decreases.

(b) The average cost will be minimized at some q for which the line through $(0, 0)$ and $(q, c(q))$ is tangent to the cost curve. This point is shown in Figure 4.31.

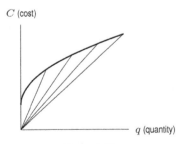

Figure 4.30

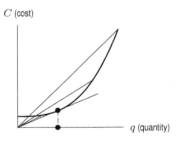

Figure 4.31

5. The cost function is $C(q) = 1000 + 20q$. The marginal cost function is the derivative $C'(q) = 20$, so the marginal cost to produce the 200th unit is $20 per unit. The average cost of producing 200 units is given by

$$a(200) = \frac{C(200)}{200} = \frac{5000}{200} = \$25/\text{unit}$$

9. (a) The average cost is $a(q) = C(q)/q$, so the total cost is $C(q) = 0.01q^3 - 0.6q^2 + 13q$.

(b) Taking the derivative of $C(q)$ gives an expression for the marginal cost:

$$C'(q) = MC(q) = 0.03q^2 - 1.2q + 13.$$

To find the smallest MC we take its derivative and find the value of q that makes it zero. So: $MC'(q) = 0.06q - 1.2 = 0$ when $q = 1.2/0.06 = 20$. This value of q must give a minimum because the graph of $MC(q)$ is a parabola opening upward. Therefore the minimum marginal cost is $MC(20) = 1$. So the marginal cost is at a minimum when the additional cost per item is $1.

(c) Differentiating gives $a'(q) = 0.02q - 0.6$.

Setting $a'(q) = 0$ and solving for q gives $q = 30$ as the quantity at which the average is minimized, since the graph of a is a parabola which opens upward. The minimum average cost is $a(30) = 4$ dollars per item.

(d) The marginal cost at $q = 30$ is $MC(30) = 0.03(30)^2 - 1.2(30) + 13 = 4$. This is the same as the average cost at this quantity. Note that since $a(q) = C(q)/q$, we have $a'(q) = (qC'(q) - C(q))/q^2$. At a critical point, q_0, of $a(q)$, we have

$$0 = a'(q_0) = \frac{q_0 C'(q_0) - C(q_0)}{q_0^2},$$

so $C'(q_0) = C(q_0)/q_0 = a(q_0)$. Therefore $C'(30) = a(30) = 4$ dollars per item.

Another way to see why the marginal cost at $q = 30$ must equal the minimum average cost $a(30) = 4$ is to view $C'(30)$ as the approximate cost of producing the 30th or 31st good. If $C'(30) < a(30)$, then producing the 31st good would lower the average cost, i.e. $a(31) < a(30)$. If $C'(30) > a(30)$, then producing the 30th good would raise the average cost, i.e. $a(30) > a(29)$. Since $a(30)$ is the global minimum, we must have $C'(30) = a(30)$.

13. Since $a(q) = C(q)/q$, we use the quotient rule to find

$$a'(q) = \frac{qC'(q) - C(q)}{q^2} = \frac{C'(q) - C(q)/q}{q} = \frac{C'(q) - a(q)}{q}.$$

Since marginal cost is C', if $C'(q) < a(q)$, then $C'(q) - a(q) < 0$, so $a'(q) < 0$.

Solutions for Section 4.6

1. The effect on the quantity demanded is approximately E times the change in price. A price increase causes a decrease in quantity demanded and a price decrease causes an increase in quantity demanded.

(a) The quantity demanded decreases by about $0.5(3\%) = 1.5\%$.

(b) The quantity demanded increases by about $0.5(3\%) = 1.5\%$.

5. Demand for high-definition TV's will be elastic, since it is not a necessary item. If the prices are too high, people will not choose to buy them, so price changes will cause relatively large demand changes.

9. The elasticity of demand for a product, E, is given by

$$E = \left| \frac{p}{q} \cdot \frac{dq}{dp} \right|.$$

We first find $dq/dp = -4p$. At a price of $5, the quantity demanded is $q = 200 - 50 = 150$ and $dq/dp = -20$, so

$$E = \left| \frac{5}{150} \cdot (-20) \right| = \frac{2}{3}.$$

Since $E < 1$ demand is inelastic.

13. (a) At a price of $2/pound, the quantity sold is

$$q = 5000 - 10(2)^2 = 5000 - 40 = 4960$$

so the total revenue is

$$R = pq = 2 \cdot 4960 = \$9{,}920$$

(b) We know that $R = pq$, and that $q = 5000 - 10p^2$, so we can substitute for q to find $R(p)$

$$R(p) = p(5000 - 10p^2) = 5000p - 10p^3$$

To find the price that maximizes revenue we take the derivative and set it equal to 0.

$$R'(p) = 0$$
$$5000 - 30p^2 = 0$$
$$30p^2 = 5000$$
$$p^2 = 166.67$$
$$p = \pm 12.91$$

We disregard the negative answer, so $p = 12.91$ is the only critical point. Is it the maximum? We use the first derivative test.

$$R'(p) > 0 \text{ if } p < 12.91 \text{ and}$$
$$R'(p) < 0 \text{ if } p > 12.91$$

So $R(p)$ has a local maximum at $p = 12.91$. We also test the function at $p = 0$, which is the only endpoint.

$$R(0) = 5000(0) - 10(0)^3 = 0$$
$$R(12.91) = 5000(12.91) - 10(12.91)^3 = 64{,}550 - 21{,}516.85 = \$43{,}033.15$$

So we conclude that revenue is maximized at price of \$12.91/pound.

(c) At a price of \$12.91/pound the quantity sold is

$$q = 5000 - 10(12.91)^2 = 5000 - 1666.68 = 3333.32$$

so the total revenue is

$$R = pq = (3333.32)(12.91) = \$43{,}033.16$$

which agrees with part (b).

(d)

$$E = \left| \frac{p}{q} \cdot \frac{dq}{dp} \right| = \left| \frac{p}{q} \cdot \frac{d}{dp}(5000 - 10p^2) \right| = \left| \frac{p}{q} \cdot (-20p) \right| = \frac{20p^2}{q}$$

Substituting $p = 12.91$ and $q = 3333.32$ yields

$$E = \frac{20(12.91)^2}{3333.32} = \frac{3333.36}{3333.32} \approx 1$$

which agrees with the result that maximum revenue occurs when $E = 1$.

17. Demand is inelastic at all prices. No matter what the price is, you can increase revenue by raising the price, so there is no actual price for which your revenue is maximized. This is not a realistic example, but it is mathematically possible. It would correspond, for instance, to the demand equation $q = 1/\sqrt{p}$, which gives revenue $R = pq = \sqrt{p}$ which is increasing for all prices $p > 0$.

21. Since marginal revenue equals dR/dq and $R = pq$, we have, using the product rule,

$$\frac{dR}{dq} = \frac{d(pq)}{dq} = p \cdot 1 + \frac{dp}{dq} \cdot q = p \left(1 + \frac{q}{p} \cdot \frac{dp}{dq} \right) = p \left(1 - \frac{1}{-\frac{p}{q} \cdot \frac{dq}{dp}} \right) = p \left(1 - \frac{1}{E} \right).$$

25. Since $R = pq$, we have $dR/dp = p(dq/dp) + q$. We are assuming that

$$0 \leq E = \left| \frac{p}{q} \cdot \frac{dq}{dp} \right| < 1$$

so, removing the absolute values

$$0 \geq -E = \frac{p}{q} \frac{dq}{dp} > -1$$

Multiplication by q gives

$$p \frac{dq}{dp} > -q$$

and hence

$$\frac{dR}{dp} = p \frac{dq}{dp} + q > 0$$

29. The approximation $E_{income} \approx \left| \frac{\Delta q/q}{\Delta I/I} \right|$ shows that the income elasticity measures the ratio of the fractional change in quantity of the product demanded to the fractional change in the income of the consumer. Thus, for example, a 1% increase in income will translate into an E_{income}% increase in the quantity purchased. After an increase in income, the consumer will tend to buy more. The income elasticity measures the strength of this tendency.

Solutions for Section 4.7

1. (a) As t gets very very large, $e^{-0.08t} \to 0$ and the function becomes $P \approx 40/1$. Thus, this model implies that when t is very large, the population is 40 billion.

(b) A graph of P against t is shown in Figure 4.32.

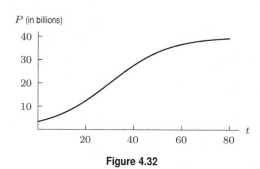

Figure 4.32

(c) We are asked to find the time t such that $P(t) = 20$. Solving we get

$$20 = P(t) = \frac{40}{1 + 11e^{-0.08t}}$$

$$1 + 11e^{-0.08t} = \frac{40}{20} = 2$$

$$11e^{-0.08t} = 1$$

$$e^{-0.08t} = \frac{1}{11}$$

$$\ln e^{-0.08t} = \ln \frac{1}{11}$$

$$-0.08t \approx -2.4$$

$$t \approx \frac{-2.4}{-0.08} = 30$$

Thus 30 years from 1990 (the year 2020) the population of the world should be 20 billion.
We are asked to find the time t such that $P(t) = 39.9$. Solving we get

$$39.9 = P(t) = \frac{40}{1 + 11e^{-0.08t}}$$

$$1 + 11e^{-0.08t} = \frac{40}{39.9} = 1.00251$$

$$11e^{-0.08t} = 0.00251$$

$$e^{-0.08t} = \frac{0.00251}{11}$$

$$\ln e^{-0.08t} = \ln \frac{0.00251}{11}$$

$$-0.08t \approx -8.39$$

$$t \approx \frac{-8.39}{-0.08} \approx 105$$

Thus 105 years from 1990 (the year 2095) the population of the world should be 39.9 billion.

5. (a), (b) The graph of $P(t)$ with carrying capacity L and point of diminishing returns t_0 is in Figure 4.33. The derivative $P'(t)$ is also shown.

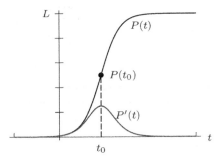

Figure 4.33

(c) Keeping track of rate of sales is the same as keeping track of the derivative $P'(t)$. The point of diminishing returns happens when the concavity of $P(t)$ changes, which is the when the derivative, $P'(t)$, switches from increasing to decreasing. This happens when $P'(t)$ reaches its maximum at $t = t_0$

9. Substituting $t = 0, 10, 20, \ldots, 70$ into the function $P = 3.9(1.03)^t$ gives the values in Table 4.1. Notice that the agreement is very close, reflecting the fact that an exponential function models the growth well over the period 1790–1860.

Table 4.1   *Predicted versus actual US population 1790–1860, in millions. (exponential model)*

Year	Actual	Predicted	Year	Actual	Predicted
1790	3.9	3.9	1830	12.9	12.7
1800	5.3	5.2	1840	17.1	17.1
1810	7.2	7.0	1850	23.2	23.0
1820	9.6	9.5	1860	31.4	30.9

13. (a) The fact that $f'(15) = 11$ means that the slope of the curve at the inflection point, $(15, 50)$ is 11. In terms of dose and response, this large slope tells us that the range of doses for which this drug is both safe and effective is small.

(b) As we can see from Figure 4.77 in the text, a dose-response curve starts out concave up (slope increasing) and switches to concave down (slope decreasing) at an inflection point. Since the slope at the inflection point $(15, 50)$ is 11, and the slope is increasing before the inflection point, $f'(10)$ is less than 11. Since the slope is decreasing after the inflection point, $f'(20)$ is also less than 11.

17. The range of safe and effective doses begins at 10 mg where the drug is effective for 100 percent of patients. It ends at 18 mg where the percent lethal curve begins to rise above zero.

Solutions for Section 4.8

1. (a) See Figure 4.34.

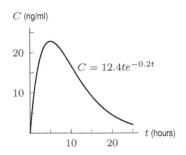

Figure 4.34

(b) The surge function $y = ate^{bt}$ changes from increasing to decreasing at $t = \frac{1}{b}$. For this function $b = 0.2$ so the peak is at $\frac{1}{0.2} = 5$ hours. We can now substitute this into the formula to compute the peak concentration:

$$C = 12.4(5)e^{-0.2(5)} = 22.8085254 \text{ ng/ml} \approx 22.8 \text{ ng/ml}.$$

(c) Tracing along the graph of $C = 12.4te^{-0.2t}$, we see it crosses the line $C = 10$ at $t \approx 1$ hour and at $t \approx 14.4$ hours. Thus, the drug is effective for $1 \leq t \leq 14.4$ hours.

(d) The drug drops below $C = 4$ for $t > 20.8$ hours. Thus, it is safe to take the other drug after 20.8 hours.

5. Large quantities of water dramatically increase the value of the peak concentration, but do not change the amount of time it takes to reach the peak concentration. The effect of the volume of water taken with the drug wears off after approximately 6 hours.

9. We find values of the parameters in the function $C = ate^{-bt}$ to create a local maximum at the point $(1.3, 23.6)$. We first set the derivative equal to zero and solve for t to find critical points. Using the product rule, we have:

$$\frac{dC}{dt} = at(e^{-bt}(-b)) + a(e^{-bt}) = 0$$
$$ae^{-bt}(-bt + 1) = 0$$
$$t = \frac{1}{b}.$$

The only critical point is at $t = 1/b$. Since we want a critical point at $t = 1.3$, we substitute and solve for b:

$$1.3 = \frac{1}{b}$$
$$b = \frac{1}{1.3} = 0.769.$$

To find the value of a, we use the fact that $C = 23.6$ when $t = 1.3$. We have:

$$a(1.3)e^{-0.769(1.3)} = 23.6$$
$$a \cdot 1.3e^{-1} = 23.6$$
$$a = \frac{23.6}{1.3e^{-1}} = 49.3.$$

We have $a = 49.3$ and $b = 0.769$.

13. (a) Products A and B have much higher peak concentrations than products C and D. Product A reaches its peak concentration slightly before products B, C, and D, which all take about the same time to reach peak concentration.

(b) If the minimum effective concentration were low, perhaps 0.2, and the maximum safe concentration were also low, perhaps 1.0, then products C and D would be the preferred drugs since they do not enter the unsafe range while being well within the effective range.

(c) If the minimum effective concentration were high, perhaps 1.2, and the maximum safe concentration were also high, perhaps 2.0, then product A would be the preferred drug since it does not enter the unsafe range and it is the only drug that is in the effective range for a substantial amount of time.

Solutions for Chapter 4 Review

1. See Figure 4.35.

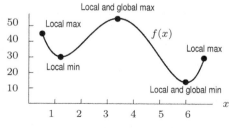

Figure 4.35

5. To find the critical points, we set $f'(x) = 0$. Since $f'(x) = 5x^4 + 60x^3$, we have

$$5x^4 + 60x^3 = 0$$
$$5x^3(x + 12) = 0$$
$$x = 0, -12.$$

There are two critical points: $x = 0$, $x = -12$.
To find the inflection points, look for points where f'' is undefined or zero. Since $f''(x) = 20x^3 + 180x^2$, it is defined everywhere. Setting it equal to zero, we get

$$20x^3 + 180x^2 = 0$$
$$20x^2(x + 9) = 0$$
$$x = 0, -9.$$

Furthermore, $20x^2(x + 9)$ is negative when $x < -9$, and positive or zero when $x > -9$. So f'' does change sign at $x = -9$, but not at $x = 0$. Thus, there is one inflection point: $x = -9$.

9. (a) $f'(x) = 1 - 1/x$. This is zero only when $x = 1$. Now $f'(x)$ is positive when $1 < x \leq 2$, and negative when $0.1 < x < 1$. Thus $f(1) = 1$ is a local minimum. The endpoints $f(0.1) \approx 2.4026$ and $f(2) \approx 1.3069$ are local maxima.

(b) Comparing values of f shows that $x = 0.1$ gives the global maximum and $x = 1$ gives the global minimum.

13. The critical points of f occur where f' is zero. These two points are indicated in the figure below.

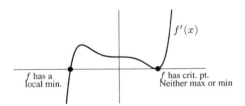

Note that the point labeled as a local minimum of f is not a critical point of f'.

17. (a) The derivative of $f(x) = x^5 + x + 7$ is $f'(x) = 5x^4 + 1$. The derivative is always positive.

(b) Since $f'(x) \neq 0$ for any value of x, there are no critical points for the function. Since $f'(x)$ is positive for all x, the function is increasing for all x, it crosses the x-axis at most once. Since $f(x) \to +\infty$ as $x \to +\infty$ and $f(x) \to -\infty$ as $x \to -\infty$, the graph of f crosses the x-axis once. So we conclude that $f(x)$ has one real root.

21. (a)

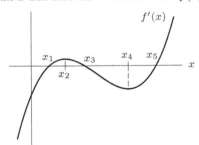

(b) $f'(x)$ changes sign at x_1, x_3, and x_5.
(c) $f'(x)$ has local extrema at x_2 and x_4.

25. See Figure 4.36.

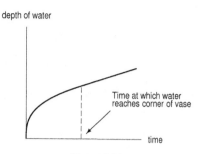

depth of water

Time at which water
reaches corner of vase

time

Figure 4.36

29. To find the intercepts of $f(x)$, we first find the y-intercept, which occurs at $f(0) = \sin(0^2) = 0$. To find the x-intercepts on the given interval, we use a calculator to graph $f(x)$ as shown in Figure 4.37:

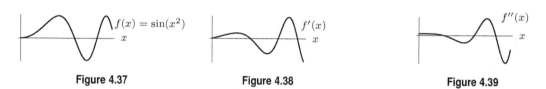

| **Figure 4.37** | **Figure 4.38** | **Figure 4.39** |

We can use a calculator's root-finding capability to find where $f(x) = 0$. We get:

$$x = 0, \quad \text{and} \quad x = 1.77, \quad \text{and} \quad x = 2.51,$$

so our intercepts are

$$(0,0), \quad (1.77, 0), \quad (2.51, 0).$$

To find critical points, we look for where $f'(x) = 0$ or f' is undefined. Using a calculator's differentiation and graphing features provides a graph of $f'(x)$ shown in Figure 4.38.

Since in Figure 4.38 all the critical points $(x, f(x))$ have $f'(x) = 0$, we use the calculator's root-finding capability to find the critical points where $f'(x) = 0$:

$$x = 0, \quad x = 1.25, \quad x = 2.17, \quad \text{and} \quad x = 2.80.$$

Writing these out as coordinates, the critical points are at

$$(0,0), \quad (1.25, 1), \quad (2.17, -1), \quad (2.80, 1).$$

Similarly, inflection points occur where $f''(x)$ changes from negative to positive or vice versa. We can look for such points on a graph of $f''(x)$, shown in Figure 4.39.

We can use the calculator's root-finding capability on the $f''(x)$ to get these inflection points:

$$x = 0.81, \quad \text{and} \quad x = 1.81, \quad \text{and} \quad x = 2.52$$

The inflection points have coordinates $(0.81, f(0.81))$, etcetera and so they are

$$(0.81, 0.61), \quad (1.81, -0.13), \quad (2.52, 0.07).$$

Notice that the intercepts can also be computed algebraically, since $\sin(x^2) = 0$ when $x^2 = 0, \pi, 2\pi$. The solutions are $x = 0$, $x = \sqrt{\pi} = 1.77$, $x = \sqrt{2\pi} = 2.51$. Similarly, if we set $f'(x) = 2x\cos(x^2) = 0$, we get $x = 0$ or $x^2 = \pi/2, 3\pi/2, 5\pi/2$. Thus, the critical points are $x = 0$, $x = \sqrt{\pi/2} = 1.25$, $x = \sqrt{3\pi/2} = 2.17$, and $x = \sqrt{5\pi/2} = 2.80$.

33. (a) We have

$$T(D) = \left(\frac{C}{2} - \frac{D}{3}\right)D^2 = \frac{CD^2}{2} - \frac{D^3}{3},$$

and

$$\frac{dT}{dD} = CD - D^2 = D(C - D).$$

Since, by this formula, dT/dD is zero when $D = 0$ or $D = C$, negative when $D > C$, and positive when $D < C$, we have (by the first derivative test) that the temperature change is maximized when $D = C$.

(b) The sensitivity is $dT/dD = CD - D^2$; its derivative is $d^2T/dD^2 = C - 2D$, which is zero if $D = C/2$, negative if $D > C/2$, and positive if $D < C/2$. Thus by the first derivative test the sensitivity is maximized at $D = C/2$.

37. The triangle in Figure 4.40 has area, A, given by

$$A = \frac{1}{2}x \cdot y = \frac{1}{2}x^3 e^{-3x}.$$

If the area has a maximum, it occurs where

$$\frac{dA}{dx} = \frac{3}{2}x^2 e^{-3x} - \frac{3}{2}x^3 e^{-3x} = 0$$

$$\frac{3}{2}x^2 (1 - x) e^{-3x} = 0$$

$$x = 0, 1.$$

The value $x = 0$ gives the minimum area, $A = 0$, for $x \geq 0$. Since

$$\frac{dA}{dx} = \frac{3}{2}x^2 (1 - x)e^{-3x},$$

we see that

$$\frac{dA}{dx} > 0 \text{ for } 0 < x < 1 \qquad \text{and} \qquad \frac{dA}{dx} < 0 \text{ for } x > 1.$$

Thus, $x = 1$ gives the local and global maximum of

$$A = \frac{1}{2}1^3 e^{-3 \cdot 1} = \frac{1}{2e^3}.$$

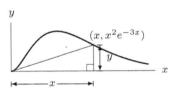

Figure 4.40

41. (a) For points (x, y) on the ellipse, we have $y^2 = 1 - x^2/9$ and $-3 \leq x \leq 3$. We wish to minimize the distance

$$D = \sqrt{(x - 2)^2 + (y - 0)^2} = \sqrt{(x - 2)^2 + 1 - \frac{x^2}{9}}.$$

To do so, we find the value of x minimizing $d = D^2$ for $-3 \leq x \leq 3$. This x also minimizes D. Since $d = (x - 2)^2 + 1 - x^2/9$, we have

$$d'(x) = 2(x - 2) - \frac{2x}{9} = \frac{16x}{9} - 4,$$

which is 0 when $x = 9/4$. Since $d''(9/4) = 16/9 > 0$, we see d has a local minimum at $x = 9/4$. Since the graph of d is a parabola, the local minimum is in fact a global minimum. Solving for y, we have

$$y^2 = 1 - \frac{x^2}{9} = 1 - \left(\frac{9}{4}\right)^2 \cdot \frac{1}{9} = \frac{7}{16},$$

so $y = \pm\sqrt{7}/4$. Therefore, the points on the ellipse closest to $(2, 0)$ are $\left(9/4, \pm\sqrt{7}/4\right)$.

(b) This time, we wish to minimize

$$D = \sqrt{(x - \sqrt{8})^2 + 1 - \frac{x^2}{9}}.$$

Again, let $d = D^2$ and minimize $d(x)$ for $-3 \le x \le 3$. Since $d = (x - \sqrt{8})^2 + 1 - x^2/9$,

$$d'(x) = 2(x - 2\sqrt{2}) - \frac{2x}{9} = \frac{16x}{9} - 4\sqrt{2}.$$

Therefore, $d'(x) = 0$ when $x = 9\sqrt{2}/4$. But $9\sqrt{2}/4 > 3$, so there are not any critical points on the interval $-3 \le x \le 3$. The minimum distance must be attained at an endpoint. Since $d'(x) < 0$ for all x between -3 and 3, the minimum is at $x = 3$. So $(3, 0)$ is the point on the ellipse closest to $(\sqrt{8}, 0)$.

45. We know that the maximum (or minimum) profit can occur when

$$\text{Marginal cost} = \text{Marginal revenue} \quad \text{or} \quad MC = MR.$$

From the table it appears that $MC = MR$ at $q \approx 2500$ and $q \approx 4500$. To decide which one corresponds to the maximum profit, look at the marginal profit at these points. Since

$$\text{Marginal profit} = \text{Marginal revenue} - \text{Marginal cost}$$

(or $M\pi = MR - MC$), we compute marginal profit at the different values of q in Table 4.2:

Table 4.2

q	1000	2000	3000	4000	5000	6000
$M\pi = MR - MC$	-22	-4	4	7	-5	-22

From the table, at $q \approx 2500$, we see that profit changes from decreasing to increasing, so $q \approx 2500$ gives a local minimum. At $q \approx 4500$, profit changes from increasing to decreasing, so $q \approx 4500$ is a local maximum. See Figure 4.41. Therefore, the global maximum occurs at $q = 4500$ or at the endpoint $q = 1000$.

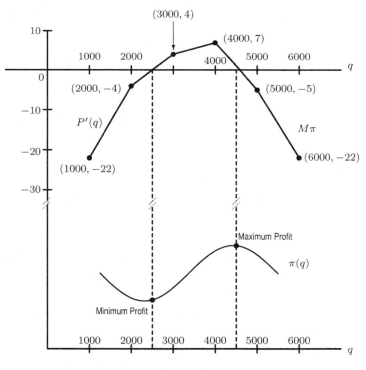

Figure 4.41

49. (a) The fixed cost is 0 because $C(0) = 0$.

(b) Profit, $\pi(q)$, is equal to money from sales, $7q$, minus total cost to produce those items, $C(q)$.

$$\pi = 7q - 0.01q^3 + 0.6q^2 - 13q$$
$$\pi' = -0.03q^2 + 1.2q - 6$$

$$\pi' = 0 \quad \text{if} \quad q = \frac{-1.2 \pm \sqrt{(1.2)^2 - 4(0.03)(6)}}{-0.06} \approx 5.9 \quad \text{or} \quad 34.1.$$

Now $\pi'' = -0.06q + 1.2$, so $\pi''(5.9) > 0$ and $\pi''(34.1) < 0$. This means $q = 5.9$ is a local min and $q = 34.1$ a local max. We now evaluate the endpoint, $\pi(0) = 0$, and the points nearest $q = 34.1$ with integer q-values:

$$\pi(35) = 7(35) - 0.01(35)^3 + 0.6(35)^2 - 13(35) = 245 - 148.75 = 96.25,$$

$$\pi(34) = 7(34) - 0.01(34)^3 + 0.6(34)^2 - 13(34) = 238 - 141.44 = 96.56.$$

So the (global) maximum profit is $\pi(34) = 96.56$. The money from sales is \$238, the cost to produce the items is \$141.44, resulting in a profit of \$96.56.

(c) The money from sales is equal to price×quantity sold. If the price is raised from \$7 by \$$x$ to \$$(7 + x)$, the result is a reduction in sales from 34 items to $(34 - 2x)$ items. So the result of raising the price by \$$x$ is to change the money from sales from $(7)(34)$ to $(7 + x)(34 - 2x)$ dollars. If the production level is fixed at 34, then the production costs are fixed at \$141.44, as found in part (b), and the profit is given by:

$$\pi(x) = (7 + x)(34 - 2x) - 141.44$$

This expression gives the profit as a function of change in price x, rather than as a function of quantity as in part (b). We set the derivative of π with respect to x equal to zero to find the change in price that maximizes the profit:

$$\frac{d\pi}{dx} = (1)(34 - 2x) + (7 + x)(-2) = 20 - 4x = 0$$

So $x = 5$, and this must give a maximum for $\pi(x)$ since the graph of π is a parabola which opens downward. The profit when the price is \$12 $(= 7 + x = 7 + 5)$ is thus $\pi(5) = (7 + 5)(34 - 2(5)) - 141.44 = \$146.56$. This is indeed higher than the profit when the price is \$7, so the smart thing to do is to raise the price by \$5.

53. If the minimum average cost occurs at a production level of 15,000 units, the line from the origin to the curve is tangent to the curve at that point. The slope of this line is 25, so the cost of producing 15,000 units is $25(15,000) = 375,000$. See Figure 4.42.

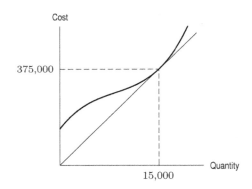

Figure 4.42

57. Both quantities are slopes of secant lines with left endpoint at $q = 50$. The value of $(C(75) - C(50))/25$ is larger as the slope of the secant line decreases as the right endpoint moves to the right.

61. $E = \left| \dfrac{p}{q} \dfrac{dq}{dp} \right| = \left| \dfrac{p}{q} \cdot \dfrac{d}{dp}(2000 - 5p) \right| = \left| \dfrac{p}{q} \cdot (-5) \right|$, so $E = \dfrac{5p}{q}$. At a price of \$20.00, the number of items produced is $q = 2000 - 5(20) = 1900$, so at $p = 20$, we have

$$E = \frac{5(20)}{(1900)} \approx 0.05.$$

Since $0 \leq E < 1$, this product has inelastic demand–a 1% change in price will only decrease demand by 0.05%.

65. The triangle in Figure 4.43 has area, A, given by

$$A = \frac{1}{2}xy = \frac{x}{2}e^{-x/3}.$$

At a critical point,

$$\frac{dA}{dx} = \frac{1}{2}e^{-x/3} - \frac{x}{6}e^{-x/3} = 0$$

$$\frac{1}{6}e^{-x/3}(3 - x) = 0$$

$$x = 3.$$

Substituting the critical point and the endpoints into the formula for the area gives:

For $x = 1$, we have $A = \frac{1}{2}e^{-1/3} = 0.358$

For $x = 3$, we have $A = \frac{3}{2}e^{-1} = 0.552$

For $x = 5$, we have $A = \frac{5}{2}e^{-5/3} = 0.472$

Thus, the maximum area is 0.552 and the minimum area is 0.358.

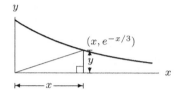

Figure 4.43

69. (a) Differentiating using the chain rule gives

$$p'(x) = \frac{1}{\sigma\sqrt{2\pi}}e^{-(x-\mu)^2/(2\sigma^2)} \cdot \left(-2\frac{(x-\mu)}{2\sigma^2}\right) = -\frac{(x-\mu)}{\sigma^3\sqrt{2\pi}}e^{-(x-\mu)^2/(2\sigma^2)}.$$

So $p'(x) = 0$ where $x - \mu = 0$, so $x = \mu$.

Differentiating again using the product rule gives

$$p''(x) = \frac{-1}{\sigma^3\sqrt{2\pi}}\left(1 \cdot e^{-(x-\mu)^2/(2\sigma^2)} + (x-\mu)e^{-(x-\mu)^2/(2\sigma^2)} \cdot \left(\frac{-2(x-\mu)}{2\sigma^2}\right)\right)$$

$$= -\frac{e^{-(x-\mu)^2/(2\sigma^2)}}{\sigma^5\sqrt{2\pi}}(\sigma^2 - (x-\mu)^2).$$

Substituting $x = \mu$ gives

$$p''(\mu) = -\frac{e^0}{\sigma^5\sqrt{2\pi}}\sigma^2 = -\frac{1}{\sigma^3\sqrt{2\pi}}.$$

Since $p''(\mu) < 0$, there is a local maximum at $x = \mu$. Since this local maximum is the only critical point, it is a global maximum. See Figure 4.44.

(b) Using the formula for $p''(x)$, we see that $p''(x) = 0$ where

$$\sigma^2 - (x-\mu)^2 = 0$$

$$x - \mu = \pm\sigma$$

$$x = \mu \pm \sigma.$$

Since $p''(x)$ changes sign at $x = \mu + \sigma$ and $x = \mu - \sigma$, these are both points of inflection. See Figure 4.44.

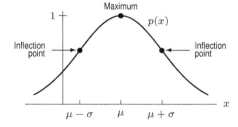

Figure 4.44

73. When bemetizide is taken with triamterene, rather than by itself, the peak concentration is lower, the time it takes to reach peak concentration is about the same, the time until onset of effectiveness is slightly less, and the duration of effectiveness is less. If bemetizide became unsafe in a concentration greater than 70 ng/ml, then it would be wise to take it with triamterene.

77. (a) To minimize A with h and s fixed we have to find $\dfrac{dA}{d\theta}$.

$$\frac{dA}{d\theta} = \frac{3}{2}s^2 \frac{d}{d\theta}\left(\frac{\sqrt{3} - \cos\theta}{\sin\theta}\right)$$

$$= \frac{3}{2}s^2 \left(\frac{\sin^2\theta - \cos\theta(\sqrt{3} - \cos\theta)}{\sin^2\theta}\right)$$

$$= \frac{3}{2}s^2 \left(\frac{\sin^2\theta - \sqrt{3}\cos\theta + \cos^2\theta}{\sin^2\theta}\right)$$

$$= \frac{3}{2}s^2 \left(\frac{1 - \sqrt{3}\cos\theta}{\sin^2\theta}\right)$$

Set $\dfrac{dA}{d\theta} = \dfrac{3}{2}s^2 \left(\dfrac{1 - \sqrt{3}\cos\theta}{\sin^2\theta}\right) = 0$. Then $1 - \sqrt{3}\cos\theta = 0$ and $\cos\theta = \frac{1}{\sqrt{3}}$, so $\theta \approx 54.7°$ is a critical point of A. Since

$$\frac{d^2 A}{d\theta^2}\bigg|_{\theta=54.7} = \frac{3}{2}s^2 \left(\frac{\sqrt{3}\sin^3\theta - 2\sin\theta\cos\theta(1 - \sqrt{3}\cos\theta)}{\sin^4\theta}\right)\bigg|_{\theta=54.7}$$

$$\approx \frac{3}{2}s^2(2.122) > 0,$$

$\theta = 54.7$ is indeed a minimum.

(b) Since $55°$ is very close to $54.7°$, we conclude that bees attempt to minimize the surface areas of their honey combs.

STRENGTHEN YOUR UNDERSTANDING

1. False. The function f has a local *minimum* at p if $f(p) \leq f(x)$ for points x near p.

5. False. If $f'(p) = 0$ and $f''(p) > 0$ then f has a local minimum at p.

9. False. Consider $f(x) = e^x$. The function and its derivative are defined for all x, and $f'(x)$ is never equal to zero. Thus, f has no critical points.

13. True. Consider $f(x) = x^3$. The point $x = 0$ is both a critical point and an inflection point of f.

17. True. For example, you can sketch the graph of such a function with a local maximum at $(-1, 1)$, a local minimum at $(1, -1)$, and horizontal asymptote $y = 0$ as $x \to \pm\infty$. For example, $f(x) = -2x/(1 + x^2)$.

21. True. Consider $f(x) = -x^2$. The function has a local and global maximum at $x = 0$.

25. False. If $f(x) < 0$ then f is decreasing on the interval, and the largest value occurs at the *left* endpoint. The global maximum will be at $x = a$.

29. True, as defined in the text.

33. False, since a point where marginal revenue equals marginal cost may also indicate minimum profit.

37. False. The cost and revenue functions cross when a profit is turning into a loss, in which case profit is decreasing, or when a loss is turning into a profit, in which case profit is increasing. In neither case is profit a maximum.

41. True, as specified in the text.

45. False. If marginal cost is less than average cost, increasing production decreases average cost.

49. False. Not necessarily. Average cost may be increasing or decreasing as a functions of quantity, depending on the concavity of the cost function.

53. False. If $0 \leq E < 1$ then we say that demand is *inelastic*.

57. False. If elasticity $E > 1$, demand is elastic and revenue increases when price decreases.

61. True. The function P fits the model $f(t) = L/(1 + Ce^{-kt})$ with $L = 1000, C = 2$, and $k = 3$.

65. True. Note that as t increases, the denominator $(1 + Ce^{-kt})$ approaches 1, so $P(t)$ approaches L.

69. True. At the inflection point, $P = L/2$, the function changes from concave up to concave down.

73. False. The function is concave down and then concave up as it approaches the t-axis asymptotically.

77. True, as specified in the text.

CHAPTER FIVE

Solutions for Section 5.1

1. (a) We compute the distances traveled for each of the three legs of the trip and add them to find the total distance traveled:

$$\text{Distance} = (30 \text{ miles/hour})(2 \text{ hours}) + (40 \text{ miles/hour})(1/2 \text{ hour}) + (20 \text{ miles/hour})(4 \text{ hours})$$
$$= 60 \text{ miles} + 20 \text{ miles} + 80 \text{ miles}$$
$$= 160 \text{ miles}.$$

You travel 160 miles on this trip.

(b) The velocity is 30 miles/hour for the first 2 hours, 40 miles/hour for the next 1/2 hour, and 20 miles/hour for the last 4 hours. The entire trip lasts $2 + 1/2 + 4 = 6.5$ hours, so we need a scale on our horizontal (time) axis running from 0 to 6.5. Between $t = 0$ and $t = 2$, the velocity is constant at 30 miles/hour, so the velocity graph is a horizontal line at 30. Likewise, between $t = 2$ and $t = 2.5$, the velocity graph is a horizontal line at 40, and between $t = 2.5$ and $t = 6.5$, the velocity graph is a horizontal line at 20. The graph is shown in Figure 5.1.

(c) How can we visualize distance traveled on the velocity graph given in Figure 5.1? The velocity graph looks like the top edges of three rectangles. The distance traveled on the first leg of the journey is (30 miles/hour)(2 hours), which is the height times the width of the first rectangle in the velocity graph. The distance traveled on the first leg of the trip is equal to the area of the first rectangle. Likewise, the distances traveled during the second and third legs of the trip are equal to the areas of the second and third rectangles in the velocity graph. It appears that distance traveled is equal to the area under the velocity graph.

In Figure 5.2, the area under the velocity graph in Figure 5.1 is shaded. Since this area is three rectangles and the area of each rectangle is given by Height × Width, we have

$$\text{Total area} = (30)(2) + (40)(1/2) + (20)(4)$$
$$= 60 + 20 + 80 = 160.$$

The area under the velocity graph is equal to distance traveled.

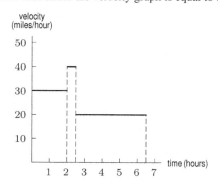

Figure 5.1: Velocity graph

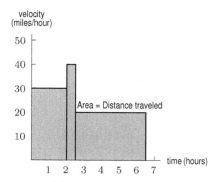

Figure 5.2: The area under the velocity graph gives distance traveled

5. We use Distance = Rate × Time on each subinterval with $\Delta t = 3$.

$$\text{Underestimate} = 0 \cdot 3 + 10 \cdot 3 + 25 \cdot 3 + 45 \cdot 3 = 240,$$
$$\text{Overestimate} = 10 \cdot 3 + 25 \cdot 3 + 45 \cdot 3 + 75 \cdot 3 = 465.$$

We know that

$$240 \leq \text{Distance traveled} \leq 465.$$

A better estimate is the average. We have

$$\text{Distance traveled} \approx \frac{240 + 465}{2} = 352.5.$$

The car travels about 352.5 feet during these 12 seconds.

9. The table gives the rate of oil consumption in billions of barrels per year. To find the total consumption, we use left-hand and right-hand sums. We have

$$\text{Left-hand sum} = (20.9)(5) + (23.3)(5) + (25.6)(5) + (28.0)(5) + (30.7)(5) = 642.5 \text{ bn barrels.}$$

$$\text{Right-hand sum} = (23.3)(5) + (25.6)(5) + (28.0)(5) + (30.7)(5) + (31.7)(5) = 696.5 \text{ bn barrels.}$$

$$\text{Average of left- and right-hand sums} = \frac{642.5 + 696.5}{2} = 669.5 \text{ bn barrels.}$$

The consumption of oil between 1985 and 2010 is about 669.5 billion barrels.

13. (a) Let's begin by graphing the data given in the table; see Figure 5.3. The total amount of pollution entering the lake during the 30-day period is equal to the shaded area. The shaded area is roughly 40% of the rectangle measuring 30 units by 35 units. Therefore, the shaded area measures about $(0.40)(30)(35) = 420$ units. Since the units are kilograms, we estimate that 420 kg of pollution have entered the lake.

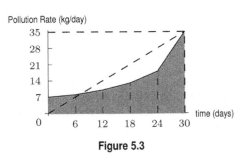

Figure 5.3

(b) Using left and right sums, we have

$$\text{Underestimate} = (7)(6) + (8)(6) + (10)(6) + (13)(6) + (18)(6) = 336 \text{ kg.}$$

$$\text{Overestimate} = (8)(6) + (10)(6) + (13)(6) + (18)(6) + (35)(6) = 504 \text{ kg.}$$

17. (a) Car A has the largest maximum velocity because the peak of car A's velocity curve is higher than the peak of B's.
(b) Car A stops first because the curve representing its velocity hits zero (on the t-axis) first.
(c) Car B travels farther because the area under car B's velocity curve is the larger.

21. (a) Based on the data, we will calculate the underestimate and the overestimate of the total change. A good estimate will be the average of both results.
Underestimate of total change

$$= 37 \cdot 10 + 41 \cdot 10 + 77 \cdot 10 + 77 \cdot 10 + 79 \cdot 10 = 3110.$$

77 was considered twice since we needed to calculate the area under the graph.
Overestimate of total change

$$= 41 \cdot 10 + 78 \cdot 10 + 78 \cdot 10 + 86 \cdot 10 + 86 \cdot 10 = 3690.$$

78 and 86 were considered twice since we needed to calculate the area over the graph.
The average is: $(3110 + 3690)/2 = 3400$ million people.
(b) The total change in the world's population between 1950 and 2000 is given by the difference between the populations in those two years. That is, the change in population equals

$$6085 \text{ (population in 2000)} - 2555 \text{ (population in 1950)} = 3530 \text{ million people.}$$

Our estimate of 3400 million people and the actual difference of 3530 million people are close to each other, suggesting our estimate was a good one.

25. To estimate the distance between the two cars at $t = 5$, we calculate upper and lower bound estimates for this distance and then average these estimates. Note that because the speed of the Prius is in miles per hour, we need to convert seconds to hours (1 sec $= 1/3600$ hours) to calculate the distance estimates in miles. We may then convert miles to feet (1 mile $= 5280$ feet).

$$\text{Distance between cars at } t = 5 \atop \text{(Upper Estimate)} \approx (\text{Difference in speeds at } t = 5) \cdot (\text{Travel time})$$

$$= (33 - 20) \cdot \frac{5}{3600} = 0.018 \text{ miles}$$

$$\text{Distance between cars at } t = 5 \atop \text{(Lower Estimate)} \approx (\text{Difference in speeds at } t = 0) \cdot (\text{Travel time})$$

$$= (0 - 0) \cdot \frac{5}{3600} = 0 \text{ miles}$$

Since:

$$\text{Average of Lower and Upper Estimates of} \atop \text{distance between cars at } t = 5 = \frac{0.018 + 0}{2} = 0.009 \text{ miles} = 48 \text{ feet,}$$

the distance between the two cars is about 48 feet, five seconds after leaving the stoplight.

29. **(a)** If $\Delta t = 4$, then $n = 2$. We have:

$$\text{Underestimate of total change } = f(0)\Delta t + f(4)\Delta t = 1 \cdot 4 + 17 \cdot 4 = 72.$$
$$\text{Overestimate of total change } = f(4)\Delta t + f(8)\Delta t = 17 \cdot 4 + 65 \cdot 4 = 328.$$

See Figure 5.4.

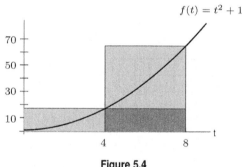

Figure 5.4

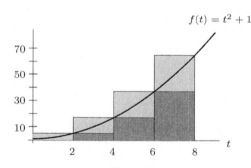

Figure 5.5

(b) If $\Delta t = 2$, then $n = 4$. We have:

$$\text{Underestimate of total change } = f(0)\Delta t + f(2)\Delta t + f(4)\Delta t + f(6)\Delta t$$
$$= 1 \cdot 2 + 5 \cdot 2 + 17 \cdot 2 + 37 \cdot 2 = 120.$$
$$\text{Overestimate of total change } = f(2)\Delta t + f(4)\Delta t + f(6)\Delta t + f(8)\Delta t$$
$$= 5 \cdot 2 + 17 \cdot 2 + 37 \cdot 2 + 65 \cdot 2 = 248.$$

See Figure 5.5.

(c) If $\Delta t = 1$, then $n = 8$.

Underestimate of total change

$$= f(0)\Delta t + f(1)\Delta t + f(2)\Delta t + f(3)\Delta t + f(4)\Delta t + f(5)\Delta t + f(6)\Delta t + f(7)\Delta t$$
$$= 1 \cdot 1 + 2 \cdot 1 + 5 \cdot 1 + 10 \cdot 1 + 17 \cdot 1 + 26 \cdot 1 + 37 \cdot 1 + 50 \cdot 1 = 148.$$

Overestimate of total change

$$= f(1)\Delta t + f(2)\Delta t + f(3)\Delta t + f(4)\Delta t + f(5)\Delta t + f(6)\Delta t + f(7)\Delta t + f(8)\Delta t$$
$$= 2 \cdot 1 + 5 \cdot 1 + 10 \cdot 1 + 17 \cdot 1 + 26 \cdot 1 + 37 \cdot 1 + 50 \cdot 1 + 65 \cdot 1 = 212.$$

See Figure 5.6.

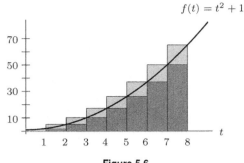

Figure 5.6

Solutions for Section 5.2

1. (a) Right sum
   (b) Upper estimate
   (c) 3
   (d) $\Delta x = 2$

5. Since e^{-x^2} is decreasing between $x = 0$ and $x = 1$, the left sum is an overestimate and the right sum is an underestimate of the integral. Letting $f(x) = e^{-x^2}$, we divide the interval $0 \le x \le 1$ into $n = 5$ sub-intervals to create Table 5.1.

Table 5.1

x	0.0	0.2	0.4	0.6	0.8	1.0
$f(x)$	1.000	0.961	0.852	0.698	0.527	0.368

(a) Letting $\Delta x = (1 - 0)/5 = 0.2$, we have:

$$\text{Left-hand sum} = f(0)\Delta x + f(0.2)\Delta x + f(0.4)\Delta x + f(0.6)\Delta x + f(0.8)\Delta x$$
$$= 1(0.2) + 0.961(0.2) + 0.852(0.2) + 0.698(0.2) + 0.527(0.2)$$
$$= 0.808.$$

(b) Again letting $\Delta x = (1 - 0)/5 = 0.2$, we have:

$$\text{Right-hand sum} = f(0.2)\Delta x + f(0.4)\Delta x + f(0.6)\Delta x + f(0.8)\Delta x + f(1)\Delta x$$
$$= 0.961(0.2) + 0.852(0.2) + 0.698(0.2) + 0.527(0.2) + 0.368(0.2)$$
$$= 0.681.$$

9. Since we are given a table of values, we must use Riemann sums to approximate the integral. Values are given every 0.2 units, so $\Delta t = 0.2$ and $n = 5$. Our best estimate is obtained by calculating the left-hand and right-hand sums, and then averaging the two.

$$\text{Left-hand sum} = 25(0.2) + 23(0.2) + 20(0.2) + 15(0.2) + 9(0.2) = 18.4$$
$$\text{Right-hand sum} = 23(0.2) + 20(0.2) + 15(0.2) + 9(0.2) + 2(0.2) = 13.8.$$

We average the two sums to obtain our best estimate of the integral:

$$\int_3^4 W(t)dt \approx \frac{18.4 + 13.8}{2} = 16.1.$$

13. $\int_0^3 f(x)\,dx$ is equal to the area shaded. We can use Riemann sums to estimate this area, or we can count grid squares. There are 3 whole grid squares and about 4 half-grid squares, for a total of 5 grid squares. Since each grid square represents 4 square units, our estimated area is $5(4) = 20$. We have $\int_0^3 f(x)\,dx \approx 20$. See Figure 5.7.

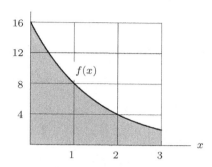

Figure 5.7

17. We have

$$3\left(\int_2^5 f(x)\,dx\right) + 1 = 3(4) + 1 = 13.$$

21. We use a calculator or computer to see that $\int_{1.1}^{1.7} e^t \ln t\,dt = 0.865$.

25. The values of the two integrals are equal, so their difference is 0.

29. **(a)** See Figure 5.8.

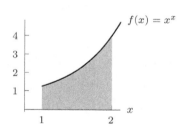

Figure 5.8

The shaded area appears to be approximately 2 units, and so $\int_1^2 x^x\,dx \approx 2$.

(b) $\int_1^2 x^x\,dx = 2.05045$

Solutions for Section 5.3

1. Since $f(x)$ is positive along the interval from 0 to 6 the area is simply $\int_0^6 (x^2 + 2)dx = 84$.

5. The two functions intersect at $x = 0$ and $x = 3$. Between these values, $3x$ is greater than x^2. We have

$$\text{Area} = \int_0^3 (3x - x^2)\,dx = 4.5.$$

9. **(a)** Counting the squares yields an estimate of 16.5, each with area = 1, so the total shaded area is approximately 16.5.

(b)

$$\int_0^8 f(x)dx = \text{(shaded area above } x\text{-axis)} - \text{(shaded area below } x\text{-axis)}$$

$$\approx 6.5 - 10 = -3.5$$

(c) The answers in (a) and (b) are different because the shaded area below the x-axis is subtracted in order to find the value of the integral in (b).

13. **(a)** The area between the graph of f and the x-axis between $x = a$ and $x = b$ is 13, so

$$\int_a^b f(x)\,dx = 13.$$

(b) Since the graph of $f(x)$ is below the x-axis for $b < x < c$,

$$\int_b^c f(x)\,dx = -2.$$

(c) Since the graph of $f(x)$ is above the x-axis for $a < x < b$ and below for $b < x < c$,

$$\int_a^c f(x)\,dx = 13 - 2 = 11.$$

(d) The graph of $|f(x)|$ is the same as the graph of $f(x)$ except that the part below the x-axis is reflected to be above it. See Figure 5.9. Thus

$$\int_a^c |f(x)|\,dx = 13 + 2 = 15.$$

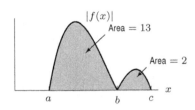

Figure 5.9

17. On the interval $0 \le x \le 5$, the entire graph lies above the x-axis, so the value of the integral is positive and equal to the area between the graph of the function and the x-axis. This area appears to be about half the area of the rectangle with area $2 \cdot 5 = 10$, so we estimate the area to be approximately 5. See Figure 5.10. Since $f(x)$ is positive on this interval, the value of the integral is equal to this area, so we have

$$\int_0^5 f(x)dx = \text{Area} \approx 5.$$

The correct match for this function is III.

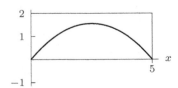

Figure 5.10

21. We can compute each integral in this problem by finding the difference between the area that lies above the x-axis and the area that lies below the x-axis on the given interval.

   (a) For $\int_0^2 f(x)\,dx$, on $0 \leq x \leq 1$ the area under the graph is 1; on $1 \leq x \leq 2$ the areas above and below the x-axis are equal and cancel each other out. Therefore, $\int_0^2 f(x)\,dx = 1$.

   (b) On $3 \leq x \leq 7$ the graph of $f(x)$ is the upper half circle of radius 2 centered at $(5,0)$. The integral is equal to the area between the graph of $f(x)$ and the x-axis, which is the area of a semicircle of radius 2. This area is 2π, and so

$$\int_3^7 f(x)\,dx = \frac{\pi 2^2}{2} = 2\pi.$$

   (c) On $2 \leq x \leq 7$ we are looking at two areas: We already know that the area of the semicircle on $3 \leq x \leq 7$ is 2π. On $2 \leq x \leq 3$, the graph lies below the x-axis and the area of the triangle is $\frac{1}{2}$. Therefore,

$$\int_2^7 f(x)\,dx = -\frac{1}{2} + 2\pi.$$

   (d) For this portion of the problem, we can split the region between the graph and the x-axis into a quarter circle on $5 \leq x \leq 7$ and a trapezoid on $7 \leq x \leq 8$ below the x-axis. The semicircle has area π, the trapezoid has area $3/2$. Therefore,

$$\int_5^8 f(x)\,dx = \pi - \frac{3}{2}.$$

25. The graph of $y = \sin x + 2$ is above the line $y = 0.5$ for $6 \leq x \leq 10$. See Figure 5.11. Therefore

$$\text{Area} = \int_6^{10} \sin x + 2 - 0.5\,dx = 7.799.$$

The integral was evaluated on a calculator.

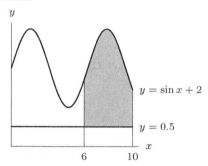

Figure 5.11

29. The graph of $y = \cos t$ is above the graph of $y = \sin t$ for $0 \leq t \leq \pi/4$ and $y = \cos t$ is below $y = \sin t$ for $\pi/4 < t < \pi$. See Figure 5.12. Therefore, we find the area in two pieces:

$$\text{Area} = \int_0^{\pi/4} (\cos t - \sin t)\,dt + \int_{\pi/4}^{\pi} (\sin t - \cos t)\,dt = 2.828.$$

The integral was evaluated on a calculator.

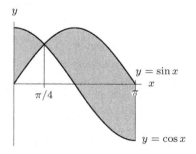

Figure 5.12

33. (a) The graph of $f(x) = x^3 - x$ crosses the x-axis at $x = 1$ since solving $f(x) = x^3 - x = 0$ gives $x = 0$ and $x = \pm 1$. See Figure 5.13. To find the total area, we find the area above the axis and the area below the axis separately. We have

$$\int_0^1 (x^3 - x)dx = -0.25 \quad \text{and} \quad \int_1^3 (x^3 - x)dx = 16.$$

As expected, the integral from 0 to 1 is negative. The area above the axis is 16 and the area below the axis is 0.25 so

$$\text{Total area} = 16.25.$$

(b) We have

$$\int_0^3 (x^3 - x)dx = 15.75.$$

Notice that the integral is equal to the area above the axis minus the area below the axis, as expected.

(c) No, they are not the same, since the integral counts area below the axis negatively while total area counts all area as positive. The two answers are not expected to be the same unless all the area lies above the horizontal axis.

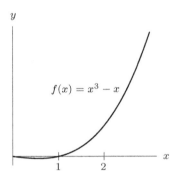

Figure 5.13

Solutions for Section 5.4

1. The integral $\int_1^3 v(t)\, dt$ represents the change in position between time $t = 1$ and $t = 3$ seconds; it is measured in meters.

5. For any t, consider the interval $[t, t + \Delta t]$. During this interval, oil is leaking out at an approximately constant rate of $f(t)$ gallons/minute. Thus, the amount of oil which has leaked out during this interval can be expressed as

$$\text{Amount of oil leaked} = \text{Rate} \times \text{Time} = f(t)\, \Delta t$$

and the units of $f(t)\, \Delta t$ are gallons/minute $\times$ minutes = gallons. The total amount of oil leaked is obtained by adding all these amounts between $t = 0$ and $t = 60$. (An hour is 60 minutes.) The sum of all these infinitesimal amounts is the integral

$$\begin{array}{c}\text{Total amount of}\\ \text{oil leaked, in gallons}\end{array} = \int_0^{60} f(t)\, dt.$$

9. (a) In 1990, when $t = 0$, gas consumption was 1770 millions of metric tons of oil equivalent. In 2010, when $t = 20$, gas consumption was $N = 1770 + 53(20) = 2830$ million metric tons of oil equivalent.

(b) We use an integral to approximate the total amount consumed over the 20-year period:

$$\text{Total amount of gas consumed} = \int_0^{20} (1770 + 53t)\, dt = 46{,}000 \text{ million metric tons of oil equivalent.}$$

13. From $t = 0$ to $t = 5$ the velocity is positive so the change in position is to the right. The area under the velocity graph gives the distance traveled. The region is a triangle, and so has area $(1/2)bh = (1/2)5 \cdot 10 = 25$. Thus the change in position is 25 cm to the right.

17. The change in the number of acres is the integral of the rate of change. We have

$$\text{Change in number of acres} = \int_0^{24} (8\sqrt{t})dt = 627.$$

The number of acres the fire covers after 24 hours is the original number of acres plus the change, so we have

$$\text{Acres covered after 24 hours} = 2000 + 627 = 2627 \text{ acres.}$$

21. The total amount of antibodies produced is

$$\text{Total antibodies} = \int_0^4 r(t)dt \approx 1.417 \text{ thousand antibodies}$$

25. (a)   (i) The income curve shows the rate of change of the value of the fund due to inflow of money. The area under the curve,

$$\int_{2000}^{2015} I(t)\, dt,$$

represents the total change in the value of the fund that is due to income. It is the quantity of money, in billions of dollars, that is projected to flow into the fund between 2000 and 2015.

(ii) The expenditure curve shows the rate of change of the value of the fund due to outflow of money. The area under the curve,

$$\int_{2000}^{2015} E(t)\, dt,$$

represents the magnitude of the change in the value of the fund that is due to expenses. It is the quantity of money, in billions of dollars, that is projected to flow out of the fund between 2000 and 2015.

(iii) The area between the income and expenditure curves,

$$\int_{2000}^{2015} I(t) - E(t)\, dt,$$

represents the difference between total income and total expenses between 2000 and 2015. It is the projected change in value of the fund between 2000 and 2015.

(b) In the figure, we see that the value of the fund was about 1000 billion dollars in 2000 and is projected to be about 3500 billion dollars in 2015. The fund is projected to increase in value by about 2500 billion dollars, and that is the area between the income and expenditure curves on the graph.

29. The area under A's curve is greater than the area under B's curve on the interval from 0 to 6, so A had the most total sales in the first 6 months. On the interval from 0 to 12, the area under B's curve is greater than the area under A, so B had the most total sales in the first year. At approximately nine months, A and B appear to have sold equal amounts. Counting the squares yields a total of about 250 sales in the first year for B and 170 sales in first year for A.

33. (a) Figure 5.14 is the graph of a rate of blood flow versus time. The total quantity of blood pumped during the three hours is given by the area under the rate graph for the three-hour time interval. The area can be estimated by counting grid boxes under the graph.

Each grid rectangle has area 30 minutes $\times$ 1/2 liter/minute $= 15$ liters, corresponding to 15 liters of blood pumped. The grid boxes in the graph are stacked in six columns. Estimating the number of boxes in each column under the graph gives

$$\text{Number of boxes} = 10 + 7.75 + 6 + 7.5 + 9 + 9.75 = 50 \text{ boxes.}$$

Approximately

$$\text{Amount of blood pumped} = 50 \cdot 15 = 750 \text{ liters.}$$

Thus, about 750 liters of blood are pumped during the three hours leading to full recovery.

(b) Since $g(t)$ is the pumping rate in liters/minute at time t hours, $60g(t)$ is the pumping rate in liters/hour. Thus $\int_0^3 60g(t)\,dt$ gives the total quantity of blood pumped in liters during the three hours.

(c) During three hours with no bleeding, the heart pumps 5 liters/minute for $3 \cdot 60 = 180$ minutes. Thus,

$$\text{Total blood pumped } = 5 \cdot 180 = 900 \text{ liters.}$$

This is $900 - 750 = 150$ liters more than pumped with 1 liter bleeding. This corresponds to the area on the graph between the 5 liters/minute line and the pumping rate for the 1 liter bleed. See Figure 5.14.

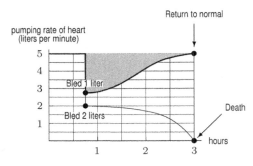

Figure 5.14

37.

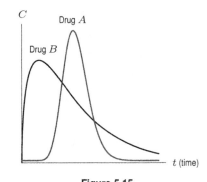

Figure 5.15

41. (a) The distance traveled is the integral of the velocity, so in T seconds you fall

$$\int_0^T 49(1 - 0.8187^t)\,dt.$$

(b) We want the number T for which

$$\int_0^T 49(1 - 0.8187^t)\,dt = 5000.$$

We can use a calculator or computer to experiment with different values for T, and we find $T \approx 107$ seconds.

Solutions for Section 5.5

1. If $H(t)$ is the temperature of the coffee at time t, by the Fundamental Theorem of Calculus

$$\text{Change in temperature } = H(10) - H(0) = \int_0^{10} H'(t)\,dt = \int_0^{10} -7(0.9^t)\,dt.$$

Therefore,

$$H(10) = H(0) + \int_0^{10} -7(0.9^t)\,dt \approx 90 - 44.2 = 45.8°\text{C.}$$

5. (a) There are approximately 5.5 squares under the curve of $C'(q)$ from 0 to 30. Each square represents $100, so the total variable cost to produce 30 units is around $550. To find the total cost, we add the fixed cost

$$\text{Total cost} = \text{fixed cost} + \text{total variable cost}$$
$$= 10,000 + 550 = \$10,550.$$

(b) There are approximately 1.5 squares under the curve of $C'(q)$ from 30 to 40. Each square represents $100, so the additional cost of producing items 31 through 40 is around $150.

(c) Examination of the graph tells us that $C'(25) = 10$. This means that the cost of producing the 26th item is approximately $10.

9. Since $C(0) = 500$, the fixed cost must be $500. The total *variable* cost to produce 20 units is

$$\int_0^{20} C'(q)\, dq = \int_0^{20} (q^2 - 16q + 70)\, dq = \$866.67 \text{ (using a calculator)}.$$

The *total* cost to produce 20 units is the fixed cost plus the variable cost of producing 20 units. Thus,

$$\text{Total cost} = \$500 + \$866.67 = \$1{,}366.67.$$

13. The total change in the net worth of the company from 2005 ($t = 0$) to 2015 ($t = 10$) is found using the Fundamental Theorem:

$$\text{Change in net worth} = f(10) - f(0) = \int_0^{10} f'(t)\, dt = \int_0^{10} (2000 - 12t^2)\, dt = 16{,}000 \text{ dollars}.$$

The worth of the company in 2015 is the worth of the company in 2005 plus the change in worth between 2005 and 2015. Thus, in 2015,

$$\text{Net worth} = f(10) = f(0) + \text{Change in worth}$$
$$= \text{Worth in 2005} + \text{Change in worth between 2005 and 2015}$$
$$= 40{,}000 + 16{,}000$$
$$= \$56{,}000.$$

17. The time period from t_0 to $2t_0$ is shorter (and contained within) the time period from 0 to $2t_0$. Thus, the amount of oil pumped out during the shorter time period, $\int_{t_0}^{2t_0} r(t)\, dt$, is less than the amount of oil pumped out in the longer timer period, $\int_0^{2t_0} r(t)\, dt$. This means

$$\int_{t_0}^{2t_0} r(t)\, dt < \int_0^{2t_0} r(t)\, dt.$$

The length of the time period from $2t_0$ to $3t_0$ is the same as the length from t_0 to $2t_0$: both are t_0 days. But the rate at which oil is pumped is going down, since $r'(t) < 0$. Thus, less oil is pumped out during the later time period, so

$$\int_{2t_0}^{3t_0} r(t)\, dt < \int_{t_0}^{2t_0} r(t)\, dt.$$

We conclude that $\int_{2t_0}^{3t_0} r(t)\, dt < \int_{t_0}^{2t_0} r(t)\, dt < \int_0^{2t_0} r(t)\, dt.$

21. The expression $\int_0^2 r(t)\, dt$ represents the amount of water that leaked from the ruptured pipe during the first two hours. Likewise, the expression $\int_2^4 r(t)\, dt$ represents the amount of water that leaked from the ruptured pipe during the next two hours. Since the leak "worsened throughout the afternoon," we know that r is an increasing function, which means that more water leaked out during the second two hours than during the first two hours. Therefore,

$$\int_0^2 r(t)\, dt < \int_2^4 r(t)\, dt.$$

Solutions for Section 5.6

1. **(a)** Since $f(x)$ is positive on the interval from 0 to 6, the integral is equal to the area under the curve. By examining the graph, we can measure and see that the area under the curve is 20 square units, so

$$\int_0^6 f(x)dx = 20.$$

(b) The average value of $f(x)$ on the interval from 0 to 6 equals the definite integral we calculated in part (a) divided by the size of the interval. Thus

$$\text{Average Value} = \frac{1}{6} \int_0^6 f(x)dx = 3\frac{1}{3}.$$

5. **(a)** Average value $= \int_0^1 \sqrt{1 - x^2}\, dx = 0.79.$

(b) The area between the graph of $y = 1 - x$ and the x-axis is 0.5. Because the graph of $y = \sqrt{1 - x^2}$ is concave down, it lies above the line $y = 1 - x$, so its average value is above 0.5. See Figure 5.16.

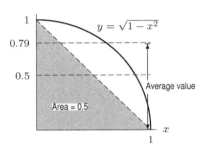

Figure 5.16

9. It appears that the area under a line at about $y = 17$ is approximately the same as the area under $f(x)$ on the interval $x = a$ to $x = b$, so we estimate that the average value is about 17. See Figure 5.17.

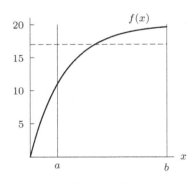

Figure 5.17

13. **(a)** Let $f(t)$ be the annual income at age t. Between ages 25 and 85,

$$\text{Average annual income} = \frac{1}{85 - 25} \int_{25}^{85} f(t)\, dt.$$

The integral equals the area of the region under the graph of $f(t)$. The region is a rectangle of height 40,000 dollars/year and width 40 years, so we have

$$\int_{25}^{85} f(t)\, dt = \text{Height} \times \text{Width} = 1{,}600{,}000 \text{ dollars}.$$

Therefore

$$\text{Average annual income} = \frac{1,600,000}{85-25} = \$26,667 \text{ per year.}$$

(b) This person spends less than their income for their entire working life, age 25 to 65. In retirement, ages 65 to 85, they have no income but can continue spending at the same rate because they saved in their working years.

17. (a) Over the interval $[-1, 3]$, we estimate that the total change of the population is about 1.5, by counting boxes between the curve and the x-axis; we count about 1.5 boxes below the x-axis from $x = -1$ to $x = 1$ and about 3 above from $x = 1$ to $x = 3$. So the average rate of change is just the total change divided by the length of the interval, that is $1.5/4 = 0.375$ thousand/hour.

(b) We can estimate the total change of the algae population by counting boxes between the curve and the x-axis. Here, there is about 1 box above the x-axis from $x = -3$ to $x = -2$, about 0.75 of a box below the x-axis from $x = -2$ to $x = -1$, and a total change of about 1.5 boxes thereafter (as discussed in part (a)). So the total change is about $1 - 0.75 + 1.5 = 1.75$ thousands of algae.

21. (a) $E(t) = 1.4e^{0.07t}$

(b)

$$\text{Average Yearly Consumption} = \frac{\text{Total Consumption for the Century}}{100 \text{ years}}$$

$$= \frac{1}{100}\int_0^{100} 1.4e^{0.07t}\,dt$$

$$\approx 219 \text{ million megawatt-hours.}$$

(c) We are looking for t such that $E(t) \approx 219$:

$$1.4e^{0.07t} \approx 219$$
$$e^{0.07t} = 156.4.$$

Taking natural logs,

$$0.07t = \ln 156.4$$
$$t \approx \frac{5.05}{0.07} \approx 72.18.$$

Thus, consumption was closest to the average during 1972.

(d) Between the years 1900 and 2000 the graph of $E(t)$ looks like

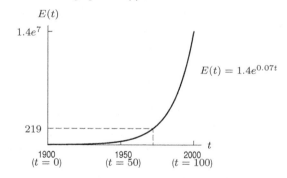

From the graph, we can see the t value such that $E(t) = 219$. It lies to the right of $t = 50$, and is thus in the second half of the century.

Solutions for Chapter 5 Review

1. (a) Since the velocity is decreasing, for an upper estimate, we use a left sum. With $n = 5$, we have $\Delta t = 2$. Then

$$\text{Upper estimate} = (44)(2) + (42)(2) + (41)(2) + (40)(2) + (37)(2) = 408.$$

(b) For a lower estimate, we use a right sum, so

$$\text{Lower estimate} = (42)(2) + (41)(2) + (40)(2) + (37)(2) + (35)(2) = 390.$$

5. (a) Right sum
   (b) Upper estimate
   (c) 4
   (d) $\Delta t = 2$
   (e) Upper estimate is approximately $12 \cdot 2 + 14.8 \cdot 2 + 16.5 \cdot 2 + 18 \cdot 2 = 122.6$.

9. (a) Using rectangles under the curve, we get

$$\text{Acres defaced} \approx (1)(0.2 + 0.4 + 1 + 2) = 3.6 \text{ acres.}$$

   (b) Using rectangles above the curve, we get

$$\text{Acres defaced} \approx (1)(0.4 + 1 + 2 + 3.5) = 6.9 \text{ acres.}$$

   (c) The number of acres defaced is between 3.6 and 6.9, so we estimate the average, 5.25 acres.

13. We use a calculator or computer to see that $\displaystyle\int_1^4 (x^2 + x)\, dx = 28.5$.

17. We use a calculator or computer to see that $\int_2^3 \frac{-1}{(r+1)^2}\, dr = -0.083$.

21. Since $x^{1/2} \le x^{1/3}$ for $0 \le x \le 1$, we have

$$\text{Area} = \int_0^1 (x^{1/3} - x^{1/2})\, dx = 0.0833.$$

   The integral was evaluated on a calculator.

25. To find the distance the car moved before stopping, we estimate the distance traveled for each two-second interval. Since speed decreases throughout, we know that the left-handed sum will be an overestimate to the distance traveled and the right-hand sum an underestimate. Applying the formulas for these sums with $\Delta t = 2$ gives:

$$\text{LEFT} = 2(100 + 80 + 50 + 25 + 10) = 530 \text{ ft.}$$
$$\text{RIGHT} = 2(80 + 50 + 25 + 10 + 0) = 330 \text{ ft.}$$

   (a) The best estimate of the distance traveled will be the average of these two estimates, or

$$\text{Best estimate} = \frac{530 + 330}{2} = 430 \text{ ft.}$$

   (b) All we can be sure of is that the distance traveled lies between the upper and lower estimates calculated above. In other words, all the black-box data tells us for sure is that the car traveled between 330 and 530 feet before stopping. So we don't know whether it hit the skunk or not; the answer is **(ii)**.

29. (a) The area under the curve is greater for species B for the first 5 years. Thus species B has a larger population after 5 years. After 10 years, the area under the graph for species B is still greater so species B has a greater population after 10 years as well.
   (b) Unless something happens that we cannot predict now, species A will have a larger population after 20 years. It looks like species A will continue to quickly increase, while species B will add only a few new plants each year.

33. The area below the x-axis is greater than the area above the x-axis, so the integral is negative.

37. The region shaded between $x = 0$ and $x = 2$ appears to have approximately the same area as the region shaded between $x = -2$ and $x = 0$, but it lies below the axis. Since $\int_{-2}^0 f(x)dx = 4$, we have the following results:
   (a) $\int_0^2 f(x)dx \approx -\int_{-2}^0 f(x)dx = -4$.
   (b) $\int_{-2}^2 f(x)dx \approx 4 - 4 = 0$.
   (c) The total area shaded is approximately $4 + 4 = 8$.

41. (a) Quantity used $= \int_0^{15} f(t)\, dt$.
   (b) Using a left hand-sum, since $\Delta t = 3$, our approximation is

$$32(1.05)^0 \cdot 3 + 32(1.05)^3 \cdot 3 + 32(1.05)^6 \cdot 3 + 32(1.05)^9 \cdot 3 + 32(1.05)^{12} \cdot 3 \approx 657.11.$$

   Since f is an increasing function, this represents an underestimate.
   (c) Each term is a lower estimate of 3 years' consumption of oil.

45. (a)

CCl$_4$ dumped

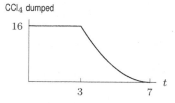

Figure 5.18

(b) 7 years, because $t^2 - 14t + 49 = (t - 7)^2$ indicates that the rate of flow was zero after 7 years.

(c)

$$\text{Area under the curve} = 3(16) + \int_3^7 (t^2 - 14t + 49)\, dt$$
$$= 69.3.$$

So $69\frac{1}{3}$ cubic yards of CCl$_4$ entered the waters from the time the EPA first learned of the situation until the flow of CCl4 was stopped.

49. Using the Fundamental Theorem of Calculus with $f(x) = 4 - x^2$ and $a = 0$, we see that

$$F(b) - F(0) = \int_0^b F'(x)dx = \int_0^b (4 - x^2)dx.$$

We know that $F(0) = 0$, so

$$F(b) = \int_0^b (4 - x^2)dx.$$

Using a calculator or computer to estimate the integral for values of b, we get

Table 5.2

b	0.0	0.5	1.0	1.5	2.0	2.5
$F(b)$	0	1.958	3.667	4.875	5.333	4.792

53. Average value $= \dfrac{1}{10 - 0} \displaystyle\int_0^{10} e^t dt \approx 2202.55$

57. It appears that the area under a line at about $y = 8.5$ is approximately the same as the area under $f(x)$ on the interval $x = a$ to $x = b$, so we estimate that the average value is about 8.5. See Figure 5.19.

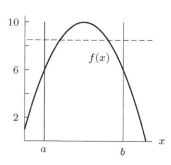

Figure 5.19

STRENGTHEN YOUR UNDERSTANDING

1. True, as specified in the text.

5. True, since the width of the interval is 2 seconds and an underestimate for the distance traveled is given by $10 \cdot 2 = 20$ ft and an overestimate for the distance traveled is given by $20 \cdot 2 = 40$ ft. The distance traveled is between 20 and 40 feet during this 2-second interval.

9. True, since the triangular area under the graph of $v(t) = 3t$ when $0 \le t \le 10$ is $(1/2)(30)(10) = 150$ feet.

13. False. Since the width of the interval is 2, the left-hand sum estimate is $2 \cdot 1000 = \$2000$.

17. False, since right-hand rectangles could be above, below, or neither. Only when f is increasing are the right-hand rectangles guaranteed to be above the graph.

21. False, since we don't know whether $f(x) \ge 0$ between $x = 0$ and $x = 2$.

25. True, since the graph of $f(x) = x^2 - 1$ is entirely on or below the x-axis between $x = -1$ and $x = 1$.

29. True, as specified in the text.

33. True, since the units of the integral are the product of the units of $a(t)$ and the units of t.

37. False. The integral gives the total *change* in the volume of water. We have to add the change to the starting volume at $t = 0$ to find the volume at $t = 30$.

41. False, since the correct statement is if $F'(t)$ is continuous, then $\int_a^b F'(t)\, dt = F(b) - F(a)$.

45. False, since by the Fundamental Theorem, $\int_0^{1000} C'(q)\, dq = C(1000) - C(0)$ which is the total *variable* costs, as we have subtracted the fixed costs of $C(0)$.

49. True, since if the total change of F over the interval $0 \le t \le 5$ is negative, then $F(0)$ must be greater than $F(5)$.

53. True, since $0 < \int_0^{10} f(x)\, dx \le \int_0^{10} 50\, dx = 500$, so the average value of f is between 0 and 50.

57. False, since, for example, $f(x)$ could be $+1$ over half of the interval $0 \le x \le 10$ and -1 over the other half.

Solutions to Problems on the Second Fundamental Theorem of Calculus

1. By the Second Fundamental Theorem, $G'(x) = x^3$.

5. (a) If $F(b) = \int_0^b 2^x\, dx$ then $F(0) = \int_0^0 2^x\, dx = 0$ since we are calculating the area under the graph of $f(x) = 2^x$ on the interval $0 \le x \le 0$, or on no interval at all.

   (b) Since $f(x) = 2^x$ is always positive, the value of F will increase as b increases. That is, as b grows larger and larger, the area under $f(x)$ on the interval from 0 to b will also grow larger.

   (c) Using a calculator or a computer, we get

$$F(1) = \int_0^1 2^x\, dx \approx 1.4,$$

$$F(2) = \int_0^2 2^x\, dx \approx 4.3,$$

$$F(3) = \int_0^3 2^x\, dx \approx 10.1.$$

9. Note that $\int_a^b (g(x))^2\, dx = \int_a^b (g(t))^2\, dt$. Thus, we have

$$\int_a^b \left((f(x))^2 - (g(x))^2\right)\, dx = \int_a^b (f(x))^2\, dx - \int_a^b (g(x))^2\, dx = 12 - 3 = 9.$$

CHAPTER SIX

Solutions for Section 6.1

1. Apply the Fundamental Theorem with $F'(x) = 2x^2 + 5$ and $a = 0$ to get values for $F(b)$. Since

$$F(b) - F(0) = \int_0^b F'(x)\,dx = \int_0^b 2x^2 + 5\,dx$$

and $F(0) = 3$, we have

$$F(b) = 3 + \int_0^b 2x^2 + 5\,dx.$$

We use a calculator or computer to estimate the definite integral $\int_0^b 2x^2 + 5\,dx$ for each value of b. For example, when $b = 0.1$, we find that $\int_0^b 2x^2 + 5\,dx = 0.501$. Thus $F(0.1) = 3.501$. Continuing in this way gives the values in Table 6.1.

Table 6.1

b	0	0.1	0.2	0.5	1.0
$F(b)$	3	3.501	4.005	5.583	6.667

5. Since $F(0) = 0$, $F(b) = \int_0^b f(t)\,dt$. For each b we determine $F(b)$ graphically as follows:
$F(0) = 0$
$F(1) = F(0) + \text{Area of } 1 \times 1 \text{ rectangle} = 0 + 1 = 1$
$F(2) = F(1) + \text{Area of triangle } (\frac{1}{2} \cdot 1 \cdot 1) = 1 + 0.5 = 1.5$
$F(3) = F(2) + \text{Negative of area of triangle} = 1.5 - 0.5 = 1$
$F(4) = F(3) + \text{Negative of area of rectangle} = 1 - 1 = 0$
$F(5) = F(4) + \text{Negative of area of rectangle} = 0 - 1 = -1$
$F(6) = F(5) + \text{Negative of area of triangle} = -1 - 0.5 = -1.5$
The graph of $F(t)$, for $0 \le t \le 6$, is shown in Figure 6.1.

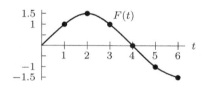

Figure 6.1

9. See Figure 6.2.

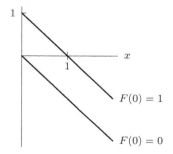

Figure 6.2

13. See Figure 6.3

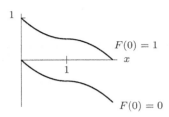

$F(0) = 1$
x
1
$F(0) = 0$

Figure 6.3

17. For every number b, the Fundamental Theorem tells us that

$$\int_0^b F'(x)\,dx = F(b) - F(0) = F(b) - 0 = F(b).$$

Therefore, the values of $F(1)$, $F(2)$, $F(3)$, and $F(4)$ are values of definite integrals. The definite integral is equal to the area of the regions under the graph above the x-axis minus the area of the regions below the x-axis above the graph. Let A_1, A_2, A_3, A_4 be the areas shown in Figure 6.4. The region between $x = 0$ and $x = 1$ lies above the x-axis, so $F(1)$ is positive, and we have

$$F(1) = \int_0^1 F'(x)\,dx = A_1.$$

The region between $x = 0$ and $x = 2$ also lies entirely above the x-axis, so $F(2)$ is positive, and we have

$$F(2) = \int_0^2 F'(x)\,dx = A_1 + A_2.$$

We see that $F(2) > F(1)$. The region between $x = 0$ and $x = 3$ includes parts above and below the x-axis. We have

$$F(3) = \int_0^3 F'(x)\,dx = (A_1 + A_2) - A_3.$$

Since the area A_3 is approximately the same as the area A_2, we have $F(3) \approx F(1)$. Finally, we see that

$$F(4) = \int_0^4 F'(x)\,dx = (A_1 + A_2) - (A_3 + A_4).$$

Since the area $A_1 + A_2$ appears to be larger than the area $A_3 + A_4$, we see that $F(4)$ is positive, but smaller than the others.

The largest value is $F(2)$ and the smallest value is $F(4)$. None of the numbers are negative.

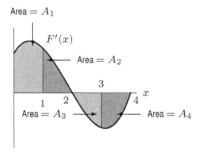

Area $= A_1$
$F'(x)$
Area $= A_2$
3
x
1 2 4
Area $= A_3$ Area $= A_4$

Figure 6.4

21. The critical points are at $(0, 5)$, $(2, 21)$, $(4, 13)$, and $(5, 15)$. A graph is given in Figure 6.5.

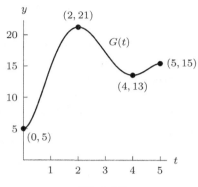

Figure 6.5

25. **(a)** Critical points of $F(x)$ are the zeros of f: $x = 1$ and $x = 3$.
  (b) $F(x)$ has a local minimum at $x = 1$ and a local maximum at $x = 3$.
  (c) See Figure 6.6.

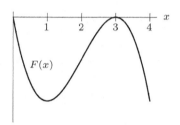

Figure 6.6

Notice that the graph could also be above or below the x-axis at $x = 3$.
29. See Figure 6.7.

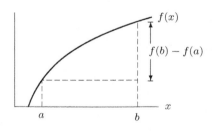

Figure 6.7

Solutions for Section 6.2

1. Since, using the chain rule,

$$F'(x) = 2e^{2x} = f(x),$$

$F(x)$ is an antiderivative of $f(x)$.

5. Since

$$F'(x) = 2\left(2e^{2x}\right) = 4e^{2x} \neq f(x),$$

$F(x)$ is not an antiderivative of $f(x)$.

9. Since 5 is a constant, an antiderivative of 5 is

$$5x + C,$$

so the expression is a family of functions.

13. The first term of the expression is a definite integral, so it is a number. The second term is an indefinite integral, hence a family of functions. Since adding a number to a family of functions gives us a new family of functions, the expression is a family of functions.

17. $5x$

21. $6(\dfrac{x^4}{4}) + 4x = \dfrac{3x^4}{2} + 4x.$

25. Antiderivative $F(x) = \dfrac{x^2}{2} + \dfrac{x^6}{6} - \dfrac{x^{-4}}{4} + C$

29. $\dfrac{t^4}{4} - \dfrac{t^3}{6} - \dfrac{t^2}{2}$

33. $F(x) = \dfrac{x^7}{7} - \dfrac{1}{7}(\dfrac{x^{-5}}{-5}) + C = \dfrac{x^7}{7} + \dfrac{1}{35}x^{-5} + C$

37. $\sin t$

41. Since

$$g'(x) = \frac{d}{dx}\left(\cos x - \sin x\right) = -\sin x - \cos x,$$

we have $g'(x) = f(x)$. So $g(x)$ is an antiderivative of $f(x)$.

45. $f(x) = 2 + 4x + 5x^2$, so $F(x) = 2x + 2x^2 + \frac{5}{3}x^3 + C$. $F(0) = 0$ implies that $C = 0$. Thus $F(x) = 2x + 2x^2 + \frac{5}{3}x^3$ is the only possibility.

49. Since $\dfrac{d}{dx}(e^x) = e^x$, we take $F(x) = e^x + C$. Now

$$F(0) = e^0 + C = 1 + C = 0,$$

so

$$C = -1$$

and

$$F(x) = e^x - 1.$$

53. $p + \ln|p| + C$

57. $5e^z + C$

61. $\dfrac{x^3}{3} + 2x^2 - 5x + C$

65. $e^x + 5x + C$

69. Since $\dfrac{d}{dt}\cos t = -\sin t$, we have

$$\int \sin t \, dt = -\cos t + C, \quad \text{where } C \text{ is a constant.}$$

73. $2\ln|x| - \pi\cos x + C$

77. $10x - 4\cos(2x) + C$

81. Since $x\sqrt{x} = x(x^{1/2}) = x^{3/2}$, an antiderivative of $x\sqrt{x}$ is:

$$\frac{x^{(3/2)+1}}{(3/2)+1} = \frac{2}{5}x^{5/2}.$$

85. (a) The marginal revenue, MR, is given by differentiating the total revenue function, R, with respect to q so

$$\frac{dR}{dq} = MR.$$

Therefore,

$$R = \int MR \, dq$$

$$= \int (20 - 4q) \, dq$$

$$= 20q - 2q^2 + C.$$

We can check this by noting

$$\frac{dR}{dq} = \frac{d}{dq}\left(20q - 2q^2 + C\right) = 20 - 4q = MR.$$

When no goods are produced the total revenue is zero so $C = 0$ and the total revenue is $R = 20q - 2q^2$.

(b) The total revenue, R, is given by pq where p is the price, so the demand curve is

$$p = \frac{R}{q} = 20 - 2q.$$

89. An antiderivative is $F(x) = -4\cos(2x) + C$. Since $F(0) = 5$, we have $5 = -4\cos 0 + C = -4 + C$, so $C = 9$. The answer is $F(x) = -4\cos(2x) + 9$.

Solutions for Section 6.3

1. Since $F'(x) = 6x$, we use $F(x) = 3x^2$. By the Fundamental Theorem, we have

$$\int_0^4 6x\,dx = 3x^2\Big|_0^4 = 3\cdot 4^2 - 3\cdot 0^2 = 48 - 0 = 48.$$

5. If $f(t) = 3t^2 + 4t + 3$, then $F(t) = t^3 + 2t^2 + 3t$. By the Fundamental Theorem, we have

$$\int_0^2 (3t^2 + 4t + 3)\,dt = (t^3 + 2t^2 + 3t)\Big|_0^2 = 2^3 + 2(2^2) + 3(2) - 0 = 22.$$

9. Since $F'(x) = 3x^2$, we take $F(x) = x^3$. Then

$$\int_0^5 3x^2\,dx = F(5) - F(0)$$

$$= 5^3 - 0^3$$

$$= 125.$$

13. Since $F'(y) = y^2 + y^4$, we take $F(y) = \frac{y^3}{3} + \frac{y^5}{5}$. Then

$$\int_0^1 (y^2 + y^4)\,dy = F(3) - F(0)$$

$$= \left(\frac{1^3}{3} + \frac{1^5}{5}\right) - \left(\frac{0^3}{3} + \frac{0^5}{5}\right)$$

$$= \frac{1}{3} + \frac{1}{5} = \frac{8}{15}.$$

17. If $F'(t) = \cos t$, we can take $F(t) = \sin t$, so

$$\int_{-1}^{1} \cos t \, dt = \sin t \Big|_{-1}^{1} = \sin 1 - \sin(-1).$$

Since $\sin(-1) = -\sin 1$, we can simplify the answer and write

$$\int_{-1}^{1} \cos t \, dt = 2 \sin 1$$

21. Since $y = x^3 - x = x(x-1)(x+1)$, the graph crosses the axis at the three points shown in Figure 6.8. The two regions have the same area (by symmetry). Since the graph is below the axis for $0 < x < 1$, we have

$$\text{Area} = 2 \left(-\int_{0}^{1} \left(x^3 - x \right) dx \right)$$

$$= -2 \left[\frac{x^4}{4} - \frac{x^2}{2} \right]_{0}^{1} = -2 \left(\frac{1}{4} - \frac{1}{2} \right) = \frac{1}{2}.$$

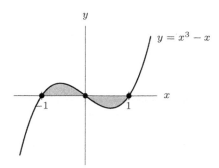

Figure 6.8

25. (a) Since r gives the rate of energy use, between 2005 and 2010 (where $t = 0$ and $t = 5$), we have

$$\text{Total energy used} = \int_{0}^{5} 462 e^{0.019t} \, dt \text{ quadrillion BTUs.}$$

(b) The Fundamental Theorem of Calculus states that

$$\int_{a}^{b} f(t) \, dt = F(b) - F(a)$$

provided that $F'(t) = f(t)$. To apply this theorem, we need to find $F(t)$ such that $F'(t) = 462 e^{0.019t}$; we take

$$F(t) = \frac{462}{0.019} e^{0.019t} = 24{,}316 e^{0.019t}.$$

Thus,

$$\text{Total energy used} = \int_{0}^{5} 462 e^{0.019t} \, dt = F(5) - F(0)$$

$$= 24{,}316 e^{0.019t} \Big|_{0}^{5}$$

$$= 24{,}316 (e^{0.095} - e^{0}) = 2423 \text{ quadrillion BTUs.}$$

Approximately 2423 quadrillion BTUs of energy were consumed between 2005 and 2010.

29. (a) An antiderivative of $F'(x) = \dfrac{1}{x^2}$ is $F(x) = -\dfrac{1}{x}$ $\left(\text{since}\dfrac{d}{dx}\left(\dfrac{-1}{x}\right) = \dfrac{1}{x^2}\right)$. So by the Fundamental Theorem we have:

$$\int_1^b \frac{1}{x^2}\,dx = -\frac{1}{x}\bigg|_1^b = -\frac{1}{b} + 1.$$

(b) Taking a limit, we have

$$\lim_{b\to\infty}\left(-\frac{1}{b} + 1\right) = 0 + 1 = 1.$$

Since the limit is 1, we know that

$$\lim_{b\to\infty}\int_1^b \frac{1}{x^2}\,dx = 1.$$

So the improper integral converges to 1:

$$\int_1^\infty \frac{1}{x^2}\,dx = 1.$$

33. (a) The total number of people that get sick is the integral of the rate. The epidemic starts at $t = 0$. Since the rate is positive for all t, we use ∞ for the upper limit of integration.

$$\text{Total number getting sick} = \int_0^\infty \left(1000te^{-0.5t}\right)\,dt$$

(b) The graph of $r = 1000te^{-0.5t}$ is shown in Figure 6.9. The shaded area represents the total number of people who get sick.

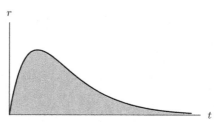

Figure 6.9

Solutions for Section 6.4

1. (a) The equilibrium price is \$30 per unit, and the equilibrium quantity is 6000.

(b) The region representing the consumer surplus is the shaded triangle in Figure 6.10 with area $\frac{1}{2}\cdot 6000\cdot 70 = 210{,}000$. The consumer surplus is \$210,000.

The area representing the producer surplus, shaded in Figure 6.11, is about 7 grid squares, each of area 10,000. The producer surplus is about \$70,000.

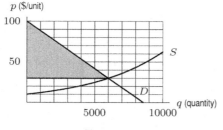

Figure 6.10

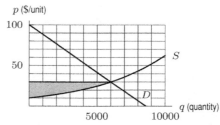

Figure 6.11

5. When $q^* = 10$, the equilibrium price is $p^* = 100 - 4 \cdot 10 = 60$. Then

$$\text{Consumer surplus} = \int_0^{10} (100 - 4q)\, dq - 60 \cdot 10.$$

Using a calculator or computer to evaluate the integral we get

$$\text{Consumer surplus} = 200.$$

Using the Fundamental Theorem of Calculus, we get:

$$
\begin{aligned}
\text{Consumer surplus} &= \int_0^{10} (100 - 4q)\, dq - 60 \cdot 10 \\
&= \left. \left(100q - 2q^2 \right) \right|_0^{10} - 600 \\
&= 1000 - 200 - 600 \\
&= 200.
\end{aligned}
$$

9. (a) Consumer surplus is greater than producer surplus in Figure 6.12.

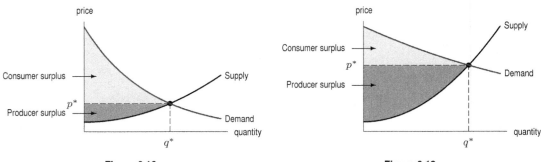

Figure 6.12   **Figure 6.13**

(b) Producer surplus is greater than consumer surplus in Figure 6.13.

13. Figure 6.14 shows the consumer and producer surplus for the price, p^-. For comparison, Figure 6.15 shows the consumer and producer surplus at the equilibrium price.

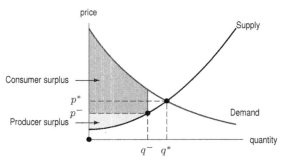
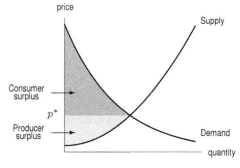

Figure 6.14   **Figure 6.15**

(a) The producer surplus is the area on the graph between p^- and the supply curve. Lowering the price also lowers the producer surplus.

(b) The consumer surplus — the area between the supply curve and the line p^- — may increase or decrease depends on the functions describing the supply and demand, and the lowered price. (For example, the consumer surplus seems to be increased in Figure 6.14 but if the price were brought down to \$0 then the consumer surplus would be zero, and hence clearly less than the consumer surplus at equilibrium.)

(c) Figure 6.14 shows that the total gains from the trade are decreased.

17.

$$\int_0^{q^*} (p^* - S(q))\,dq = \int_0^{q^*} p^*\,dq - \int_0^{q^*} S(q)\,dq$$

$$= p^*q^* - \int_0^{q^*} S(q)\,dq.$$

Using Problem 16, this integral is the extra amount consumers pay (i.e., suppliers earn over and above the minimum they would be willing to accept for supplying the good). It results from charging the equilibrium price.

Solutions for Section 6.5

1.

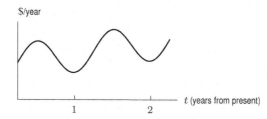

$/year

t (years from present)

The graph reaches a peak each summer, and a trough each winter. The graph shows sunscreen sales increasing from cycle to cycle. This gradual increase may be due in part to inflation and to population growth.

5. (a) The future value is
$$\text{Future value } = 12{,}000e^{0.05(6)} = \$16{,}198.31.$$

(b) We find the present value of the income stream first. Using a calculator or computer to evaluate the definite integral, we get
$$\text{Present value } = \int_0^6 2000e^{-0.05t}\,dt = \$10{,}367.27.$$
Using instead the Fundamental Theorem of Calculus, we get:
$$\text{Present value} = \int_0^6 2000e^{-0.05t}\,dt$$
$$= 2000\left(\frac{1}{-0.05}\right)e^{-0.05t}\Big|_0^6$$
$$= -40000\left(e^{-0.3} - e^0\right) = \$10{,}367.27.$$
We use the present value to find the future value.
$$\text{Future value } = 10{,}367.27e^{0.05(6)} = \$13{,}994.35.$$

(c) Although we deposit the exact same amount in the two situations, the future value is larger for the lump sum. It is always financially preferable to receive the money earlier rather than later, since it has more time to earn interest.

9. (a) (i) Using a calculator or computer with an interest rate of 3%, we have
$$\text{Present value } = \int_0^4 5000e^{-0.03t}\,dt = \$18{,}846.59.$$
Using instead the Fundamental Theorem of Calculus, we get:
$$P = \int_0^4 5000e^{-0.03t}\,dt$$
$$= 5000\left(\frac{1}{-0.03}\right)e^{-0.03t}\Big|_0^4$$
$$= -166666.67\left(e^{-0.12} - e^0\right) = \$18{,}846.59$$

(ii) If the interest rate is 10%, using a calculator or computer we get

$$\text{Present value} = \int_0^4 5000e^{-0.10t}\, dt = \$16,484.00.$$

Using instead the Fundamental Theorem of Calculus, we get:

$$P = \int_0^4 5000e^{-0.10t}\, dt$$

$$= 5000 \left(\frac{1}{-0.10}\right) e^{-0.10t} \Big|_0^4$$

$$= -50000 \left(e^{-0.4} - e^0\right) = \$16,484.00$$

(b) At the end of the four-year period, if the interest rate is 3%,

$$\text{Value} = 18,846.59e^{0.03(4)} = \$21,249.47.$$

At 10%,

$$\text{Value} = 16,484.00e^{0.10(4)} = \$24,591.24.$$

13. (a) The future value in 10 years is $100,000. We first find the present value, P:

$$100000 = Pe^{0.10(10)}$$

$$P = \$36,787.94$$

We solve for the income stream S, using a calculator or computer to evaluate the definite integral:

$$36,787.94 = \int_0^{10} Se^{-0.10t}\, dt$$

$$36,787.94 = S \int_0^{10} e^{-0.10t}\, dt$$

$$36,787.94 = S(6.321)$$

Using instead the Fundamental Theorem of Calculus to evaluate the definite integral, we get

$$\int_0^{10} e^{-0.10t}\, dt = \left(\frac{1}{-0.10}\right) e^{-0.10t} \Big|_0^{10} = 6.321.$$

Solving for S we get

$$S = \frac{36,787.94}{6.321} = \$5820.00 \text{ per year.}$$

The income stream required is about $5820 per year (or about $112 per week).

(b) The present value is $36,787.94. This is the amount that would have to be deposited now.

17. (a) The income stream is $34.6 billion per year and the interest rate is 6%. Using a calculator or computer to evaluate the integral, we get

$$\text{Present value} = \int_0^1 34.6e^{-0.06t}\, dt$$

$$= 33.58 \text{ billion dollars.}$$

Using instead the Fundamental Theorem of Calculus, we get:

$$\text{Present value} = \int_0^1 34.6e^{-0.06t}\, dt$$

$$= 34.6 \left(\frac{1}{-0.06}\right) e^{-0.06t} \Big|_0^1$$

$$= -(576.667) \left(e^{-0.06} - e^0\right) = \$33.58 \text{ billion dollars.}$$

The present value of Intel's profits over the one-year time period is about 33.38 billion dollars.

(b) The value at the end of the year is $33.58e^{0.06(1)} = 35.66$, or about 35.66 billion dollars.

21. (a) Suppose the oil extracted over the time period $[0, M]$ is S. (See Figure 6.16.) Since $q(t)$ is the rate of oil extraction, we have:

$$S = \int_0^M q(t)dt = \int_0^M (a - bt)dt = \int_0^M (10 - 0.1t)\, dt.$$

To calculate the time at which the oil is exhausted, set $S = 100$ and try different values of M. We find $M = 10.6$ gives

$$\int_0^{10.6} (10 - 0.1t)\, dt = 100,$$

so the oil is exhausted in 10.6 years.

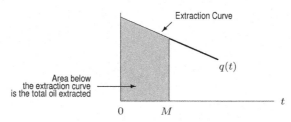

Figure 6.16

(b) Suppose p is the oil price, C is the extraction cost per barrel, and r is the interest rate. We have the present value of the profit as

$$\begin{aligned}
\text{Present value of profit} &= \int_0^M (p - C)q(t)e^{-rt}dt \\
&= \int_0^{10.6} (20 - 10)(10 - 0.1t)e^{-0.1t}\, dt \\
&= 624.9 \text{ million dollars.}
\end{aligned}$$

Solutions for Section 6.6

1. (a) $\frac{d}{dx}\sin(x^2 + 1) = 2x\cos(x^2 + 1)$; $\quad \frac{d}{dx}\sin(x^3 + 1) = 3x^2\cos(x^3 + 1)$
   (b) (i) $\frac{1}{2}\sin(x^2 + 1) + C$ $\quad$ (ii) $\frac{1}{3}\sin(x^3 + 1) + C$
   (c) (i) $-\frac{1}{2}\cos(x^2 + 1) + C$ $\quad$ (ii) $-\frac{1}{3}\cos(x^3 + 1) + C$

5. Setting $w = \sin t$, we see that $dw = \cos t\, dt$. Since the integrand has a factor of $\cos t$ and no additional factors, substitution is appropriate and allows us to replace the integrand by one we can integrate directly using the power rule:

$$\int \sin^9 t\, (\cos t\, dt) = \int w^9\, dw.$$

9. We use the substitution $w = x^2 + 1$, $dw = 2x dx$.

$$\int x\sqrt{x^2 + 1}dx = \frac{1}{2}\int w^{1/2}dw = \frac{1}{3}w^{3/2} + C = \frac{1}{3}(x^2 + 1)^{3/2} + C.$$

Check: $\frac{d}{dx}\left(\frac{1}{3}(x^2 + 1)^{3/2} + C\right) = \frac{1}{3}\cdot\frac{3}{2}(x^2 + 1)^{1/2}\cdot 2x = x\sqrt{x^2 + 1}.$

13. We use the substitution $w = -0.2t$, $dw = -0.2dt$.

$$\int 100e^{-0.2t}dt = \frac{100}{-0.2}\int e^w dw = -500e^w + C = -500e^{-0.2t} + C.$$

17. We use the substitution $w = x^2 + 3$, $dw = 2x\,dx$.

$$\int x(x^2+3)^2\,dx = \int w^2\left(\frac{1}{2}\,dw\right) = \frac{1}{2}\frac{w^3}{3} + C = \frac{1}{6}(x^2+3)^3 + C.$$

Check: $\dfrac{d}{dx}\left[\dfrac{1}{6}(x^2+3)^3 + C\right] = \dfrac{1}{6}\left[3(x^2+3)^2(2x)\right] = x(x^2+3)^2.$

21. We use the substitution $w = 2t - 7$, $dw = 2\,dt$.

$$\int (2t-7)^{73}\,dt = \frac{1}{2}\int w^{73}\,dw = \frac{1}{(2)(74)}w^{74} + C = \frac{1}{148}(2t-7)^{74} + C.$$

Check: $\dfrac{d}{dt}\left[\dfrac{1}{148}(2t-7)^{74} + C\right] = \dfrac{74}{148}(2t-7)^{73}(2) = (2t-7)^{73}.$

25. We use the substitution $w = \sin\theta$, $dw = \cos\theta\,d\theta$.

$$\int \sin^6\theta\cos\theta\,d\theta = \int w^6\,dw = \frac{w^7}{7} + C = \frac{\sin^7\theta}{7} + C.$$

Check: $\dfrac{d}{d\theta}\left[\dfrac{\sin^7\theta}{7} + C\right] = \sin^6\theta\cos\theta.$

29. We use the substitution $w = x^3 + 1$, $dw = 3x^2\,dx$, to get

$$\int x^2 e^{x^3+1}\,dx = \frac{1}{3}\int e^w\,dw = \frac{1}{3}e^w + C = \frac{1}{3}e^{x^3+1} + C.$$

Check: $\dfrac{d}{dx}\left(\dfrac{1}{3}e^{x^3+1} + C\right) = \dfrac{1}{3}e^{x^3+1}\cdot 3x^2 = x^2 e^{x^3+1}.$

33. We use the substitution $w = 3x - 4$, $dw = 3dx$.

$$\int e^{3x-4}\,dx = \frac{1}{3}\int e^w\,dw = \frac{1}{3}e^w + C = \frac{1}{3}e^{3x-4} + C.$$

Check: $\dfrac{d}{dx}\left(\dfrac{1}{3}e^{3x-4} + C\right) = \dfrac{1}{3}e^{3x-4}\cdot 3 = e^{3x-4}.$

37. We use the substitution $w = e^x + e^{-x}$, $dw = (e^x - e^{-x})\,dx$.

$$\int \frac{e^x - e^{-x}}{e^x + e^{-x}}\,dx = \int \frac{dw}{w} = \ln|w| + C = \ln(e^x + e^{-x}) + C.$$

(We can drop the absolute value signs since $e^x + e^{-x} > 0$ for all x).

Check: $\dfrac{d}{dx}[\ln(e^x + e^{-x}) + C] = \dfrac{1}{e^x + e^{-x}}(e^x - e^{-x}).$

41. We use the substitution $w = e^t + t$, $dw = (e^t + 1)\,dt$.

$$\int \frac{e^t + 1}{e^t + t}\,dt = \int \frac{1}{w}\,dw = \ln|w| + C = \ln|e^t + t| + C.$$

Check: $\dfrac{d}{dt}(\ln|e^t + t| + C) = \dfrac{e^t + 1}{e^t + t}.$

45. We use the substitution $w = e^t + 1$, $dw = e^t dt$:

$$\int \frac{e^t}{e^t + 1}\,dt = \int \frac{1}{w}\,dw = \ln|w| + C = \ln(e^t + 1) + C.$$

49. (a) We substitute $w = 1 + x^2$, $dw = 2x\,dx$.

$$\int_{x=0}^{x=1} \frac{x}{1+x^2}\,dx = \frac{1}{2}\int_{w=1}^{w=2}\frac{1}{w}\,dw = \frac{1}{2}\ln|w|\Big|_1^2 = \frac{1}{2}\ln 2.$$

(b) We substitute $w = \cos x$, $dw = -\sin x\, dx$.

$$\int_{x=0}^{x=\frac{\pi}{4}} \frac{\sin x}{\cos x}\, dx = -\int_{w=1}^{w=\sqrt{2}/2} \frac{1}{w}\, dw$$

$$= -\ln|w| \Big|_{1}^{\sqrt{2}/2} = -\ln\frac{\sqrt{2}}{2} = \frac{1}{2}\ln 2.$$

53. Let $\sqrt{x} = w$, $\frac{1}{2}x^{-\frac{1}{2}}\, dx = dw$, $\frac{dx}{\sqrt{x}} = 2\, dw$. If $x = 1$ then $w = 1$, and if $x = 4$ so $w = 2$. So we have

$$\int_{1}^{4} \frac{e^{\sqrt{x}}}{\sqrt{x}}\, dx = \int_{1}^{2} e^{w} \cdot 2\, dw = 2e^{w} \Big|_{1}^{2} = 2(e^2 - e) \approx 9.34.$$

57. Let $w = -t^2$, then $dw = -2t\,dt$ so $t\,dt = -\frac{1}{2}dw$. When $t = 0, w = 0$ and when $t = 1, w = -1$. Thus we have

$$\int_{0}^{1} 2te^{-t^2}\, dt = \int_{0}^{-1} 2e^{w}\left(-\frac{1}{2}dw\right) = -\int_{0}^{-1} e^{w}\, dw$$

$$= -e^{w} \Big|_{0}^{-1} = -e^{-1} - (-e^0) = 1 - e^{-1}.$$

61. Since $f(x) = 1/(x+1)$ is positive on the interval $x = 0$ to $x = 2$, we have

$$\text{Area} = \int_{0}^{2} \frac{1}{x+1}dx = \ln(x+1) \Big|_{0}^{2} = \ln 3 - \ln 1 = \ln 3.$$

The area is $\ln 3 \approx 1.0986$.

65. (a) $\displaystyle\int 4x(x^2 + 1)\, dx = \int (4x^3 + 4x)\, dx = x^4 + 2x^2 + C.$

(b) If $w = x^2 + 1$, then $dw = 2x\, dx$.

$$\int 4x(x^2 + 1)\, dx = \int 2w\, dw = w^2 + C = (x^2 + 1)^2 + C.$$

(c) The expressions from parts (a) and (b) look different, but they are both correct. Note that $(x^2 + 1)^2 + C = x^4 + 2x^2 + 1 + C$. In other words, the expressions from parts (a) and (b) differ only by a constant, so they are both correct antiderivatives.

Solutions for Section 6.7

1. Let $u = t$ and $v' = e^{5t}$, so $u' = 1$ and $v = \frac{1}{5}e^{5t}$.
Then $\int te^{5t}\, dt = \frac{1}{5}te^{5t} - \int \frac{1}{5}e^{5t}\, dt = \frac{1}{5}te^{5t} - \frac{1}{25}e^{5t} + C.$

5. Let $u = \ln 5q$, $v' = q^5$. Then $v = \frac{1}{6}q^6$ and $u' = \dfrac{1}{q}$. Integrating by parts, we get:

$$\int q^5 \ln 5q\, dq = \frac{1}{6}q^6 \ln 5q - \int \left(5 \cdot \frac{1}{5q}\right) \cdot \frac{1}{6}q^6\, dq$$

$$= \frac{1}{6}q^6 \ln 5q - \frac{1}{36}q^6 + C.$$

9. Let $u = y$ and $v' = \frac{1}{\sqrt{5-y}}$, so $u' = 1$ and $v = -2(5-y)^{1/2}$.

$$\int \frac{y}{\sqrt{5-y}}\, dy = -2y(5-y)^{1/2} + 2\int (5-y)^{1/2}\, dy = -2y(5-y)^{1/2} - \frac{4}{3}(5-y)^{3/2} + C.$$

13. Let $u = t$, $v' = \sin t$. Thus, $v = -\cos t$ and $u' = 1$. With this choice of u and v, integration by parts gives:

$$\int t \sin t \, dt = -t \cos t - \int (-\cos t) \, dt$$
$$= -t \cos t + \sin t + C.$$

17. $\displaystyle\int_1^5 \ln t \, dt = (t \ln t - t)\Big|_1^5 = 5 \ln 5 - 4 \approx 4.047$

21. Let $u = (\ln t)^2$ and $v' = 1$, so $u' = \dfrac{2 \ln t}{t}$ and $v = t$. Then

$$\int (\ln t)^2 \, dt = t(\ln t)^2 - 2 \int \ln t \, dt = t(\ln t)^2 - 2t \ln t + 2t + C.$$

(We use the fact that $\displaystyle\int \ln x \, dx = x \ln x - x + C$, a result which can be derived using integration by parts.)

25. Since $\ln(x^2) = 2 \ln x$ and $\int \ln x \, dx = x \ln x - x + C$, we have

$$\text{Area} = \int_1^2 (\ln(x^2) - \ln x) \, dx = \int_1^2 (2 \ln x - \ln x) \, dx$$
$$= \int_1^2 \ln x \, dx = (x \ln x - x)\Big|_1^2 = 2 \ln 2 - 2 - (1 \ln 1 - 1) = 2 \ln 2 - 1.$$

Solutions for Chapter 6 Review

1. Since dP/dt is negative for $t < 3$ and positive for $t > 3$, we know that P is decreasing for $t < 3$ and increasing for $t > 3$. Between each two integer values, the magnitude of the change is equal to the area between the graph dP/dt and the t-axis. For example, between $t = 0$ and $t = 1$, we see that the change in P is -1. Since $P = 2$ at $t = 0$, we must have $P = 1$ at $t = 1$. The other values are found similarly, and are shown in Table 6.2.

Table 6.2

t	1	2	3	4	5
P	1	0	$-1/2$	0	1

5. (a) The function $f(x)$ is increasing when $f'(x)$ is positive, so $f(x)$ is increasing for $x < -2$ or $x > 2$.
The function $f(x)$ is decreasing when $f'(x)$ is negative, so $f(x)$ is decreasing for $-2 < x < 2$.
Since $f(x)$ is increasing to the left of $x = -2$, decreasing between $x = -2$ and $x = 2$, and increasing to the right of $x = 2$, the function $f(x)$ has a local maximum at $x = -2$ and a local minimum at $x = 2$.
(b) See Figure 6.17.

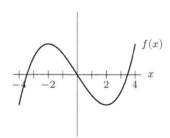

Figure 6.17

9. $t^3 + \dfrac{7t^2}{2} + t.$

13. $x^3 + 5x.$

17. We use substitution with $w = 2x + 1$ and $dw = 2\,dx$. Then

$$\int f(x)\,dx = \int (2x+1)^3\,dx = \int w^3 \frac{1}{2}\,dw = \frac{w^4}{2 \cdot 4} + C = \frac{(2x+1)^4}{8} + C.$$

21. $\dfrac{x^4}{4} - \dfrac{x^2}{2} + C.$

25. $2x^2 + 2e^x + C$

29. Since $f(x) = \dfrac{x+1}{x} = 1 + \dfrac{1}{x}$, the indefinite integral is $x + \ln|x| + C$

33. $\displaystyle\int_0^1 \sin\theta\,d\theta = -\cos\theta\,\Big|_0^1 = 1 - \cos 1 \approx 0.460.$

37. The integral which represents the area under this curve is

$$\text{Area} = \int_0^2 (6x^2 + 1)\,dx.$$

Since $\dfrac{d}{dx}(2x^3 + x) = 6x^2 + 1$, we can evaluate the definite integral:

$$\int_0^2 (6x^2 + 1)\,dx = (2x^3 + x)\,\Big|_0^2 = 2(2^3) + 2 - (2(0) + 0) = 16 + 2 = 18.$$

41. Since $y = 0$ only when $x = 0$ and $x = 1$, the area lies between these limits and is given by

$$\text{Area} = \int_0^1 x^2(1-x)^2\,dx = \int_0^1 x^2(1 - 2x + x^2)\,dx = \int_0^1 (x^2 - 2x^3 + x^4)\,dx$$

$$= \frac{x^3}{3} - \frac{2}{4}x^4 + \frac{x^5}{5}\,\Big|_0^1 = \frac{1}{30}.$$

45. **(a)** Since $v(t) = 60/50^t$ is never 0, the car never stops.
(b) For time $t \geq 0$,

$$\text{Distance traveled} = \int_0^\infty \frac{60}{50^t}\,dt.$$

(c) Evaluating $\displaystyle\int_0^b \frac{60}{50^t}\,dt$ for $b = 1, 5, 10$ gives

$$\int_0^1 \frac{60}{50^t}\,dt = 15.0306 \qquad \int_0^5 \frac{60}{50^t}\,dt = 15.3373 \qquad \int_0^{10} \frac{60}{50^t}\,dt = 15.3373,$$

so the integral appears to converge to 15.3373; so we estimate the distance traveled to be 15.34 miles.

49. Measuring money in thousands of dollars, the equation of the line representing the demand curve passes through (50, 980) and (350, 560). Its slope is $(560 - 980)/(350 - 50) = -420/300$. See Figure 6.18. So the equation is $y - 560 = -\frac{420}{300}(x - 350)$, i.e. $y - 560 = -\frac{7}{5}x + 490$. Thus

$$\text{Consumer surplus} = \int_0^{350} \left(-\frac{7}{5}x + 1050\right) dx - 350 \cdot 560 = 85{,}750.$$

(Note that $85{,}750 = \frac{1}{2} \cdot 490 \cdot 350$, the area of the triangle in Figure 6.18. We could have used this instead of the integral to find the consumer surplus.)

Recalling that our unit measure for the price axis is \$1000/car, the consumer surplus is \$85,750,000.

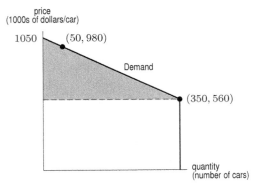

price
(1000s of dollars/car)

1050

(50, 980)

Demand

(350, 560)

quantity
(number of cars)

Figure 6.18

53.

$$\text{Present Value} = \int_0^{10} (100 + 10t)e^{-.05t} \, dt$$
$$= \$1{,}147.75.$$

57. $f(x) = -7x$, so $F(x) = \frac{-7x^2}{2} + C$. $F(0) = 0$ implies that $-\frac{7}{2} \cdot 0^2 + C = 0$, so $C = 0$. Thus $F(x) = -7x^2/2$ is the only possibility.

61. We use the substitution $w = x^2 + 1$, $dw = 2x \, dx$.

$$\int \frac{2x}{x^2 + 1} \, dx = \int \frac{1}{w} \, dw = \ln|w| + C = \ln(x^2 + 1) + C.$$

65. We use the substitution $w = -x$, $dw = -dx$.

$$\int e^{-x} \, dx = -\int e^w \, dw = -e^w + C = -e^{-x} + C.$$

Check: $\frac{d}{dx}(-e^{-x} + C) = -(-e^{-x}) = e^{-x}$.

69. Since $f'(x) = 2x$, integration by parts tells us that

$$\int_0^{10} f(x)g'(x) \, dx = f(x)g(x) \Big|_0^{10} - \int_0^{10} f'(x)g(x) \, dx$$
$$= f(10)g(10) - f(0)g(0) - 2\int_0^{10} xg(x) \, dx.$$

We can use left and right Riemann Sums with $\Delta x = 2$ to approximate $\int_0^{10} xg(x) \, dx$:

$$\text{Left sum} \approx 0 \cdot g(0)\Delta x + 2 \cdot g(2)\Delta x + 4 \cdot g(4)\Delta x + 6 \cdot g(6)\Delta x + 8 \cdot g(8)\Delta x$$
$$= (0(2.3) + 2(3.1) + 4(4.1) + 6(5.5) + 8(5.9)) \, 2 = 205.6.$$
$$\text{Right sum} \approx 2 \cdot g(2)\Delta x + 4 \cdot g(4)\Delta x + 6 \cdot g(6)\Delta x + 8 \cdot g(8)\Delta x + 10 \cdot g(10)\Delta x$$
$$= (2(3.1) + 4(4.1) + 6(5.5) + 8(5.9) + 10(6.1)) \, 2 = 327.6.$$

A good estimate for the integral is the average of the left and right sums, so

$$\int_0^{10} xg(x) \, dx \approx \frac{205.6 + 327.6}{2} = 266.6.$$

Substituting values for f and g, we have

$$\int_0^{10} f(x)g'(x) \, dx = f(10)g(10) - f(0)g(0) - 2\int_0^{10} xg(x) \, dx$$
$$\approx 10^2(6.1) - 0^2(2.3) - 2(266.6) = 76.8 \approx 77.$$

73. Remember that $\ln(x^2) = 2 \ln x$. Therefore,

$$\int \ln(x^2)\, dx = 2 \int \ln x\, dx = 2x \ln x - 2x + C.$$

Check:

$$\frac{d}{dx}(2x \ln x - 2x + C) = 2 \ln x + \frac{2x}{x} - 2 = 2 \ln x = \ln(x^2).$$

STRENGTHEN YOUR UNDERSTANDING

1. True.

5. True. Since f' is positive on the interval 3 to 4, the function is increasing on that interval.

9. False. Since f' is negative on the interval 0 to 1, the function is decreasing on that interval.

13. False. Antiderivatives of e^{3x} are of the form $(1/3)e^{3x} + C$.

17. False. The derivative of $\ln|t|$ is $1/t$ so the correct integral statement is $\int (1/t)\, dt = \ln|t| + C$.

21. False. We need to substitute the endpoints into an antiderivative of $1/x$.

25. False. For a definite integral, we need to substitute the endpoints into the antiderivative.

29. False. The two definite integrals represent two different quantities.

33. False. The equilibrium price is the price where the supply curve crosses the demand curve.

37. False. Total gains from trade is the *sum* of consumer and producer surplus.

41. True. In M years, future value = present value $\cdot e^{rM}$, and since the interest rate r is positive, the future value is greater.

45. False. An income stream has units of dollars/year.

49. False. The present value of an income stream of 2000 dollars per year that starts now and pays out over 6 years with a continuous interest rate of 2% is $\int_0^6 2000e^{-.02t}\, dt$.

53. False. We have $dw = 2x\, dx$. Since the integral $\int e^{x^2}\, dx$ does not include an $x\, dx$ to be substituted for dw, this integral cannot be evaluated using this substitution.

57. True, since $dw = (e^x - e^{-x})\, dx$.

61. False. The integral has to be $\int u\, dv$ when u and dv are substituted. In this case, we should have $u = x^2$ and $dv = e^x\, dx$.

65. False. This integral is more appropriately evaluated using the method of substitution.

69. False. This integral is more appropriately evaluated using the method of substitution.

FOCUS ON PRACTICE

1. We use the substitution $w = y^2 + 5$, $dw = 2y\,dy$.

$$\int y(y^2 + 5)^8\, dy = \frac{1}{2} \int (y^2 + 5)^8 (2y\,dy)$$
$$= \frac{1}{2} \int w^8\, dw = \frac{1}{2} \frac{w^9}{9} + C$$
$$= \frac{1}{18}(y^2 + 5)^9 + C.$$

Check: $\dfrac{d}{dy}\left(\dfrac{1}{18}(y^2 + 5)^9 + C\right) = \dfrac{1}{18}[9(y^2 + 5)^8(2y)] = y(y^2 + 5)^8.$

5. Since $\dfrac{d}{dx}(e^{-3t}) = -3e^{-3t}$, we have $\displaystyle\int e^{-3t}\, dt = -\dfrac{1}{3}e^{-3t} + C.$

9. $\dfrac{x^4}{4} + 2x^2 + 8x + C.$

13. $\dfrac{q^3}{3} + \dfrac{5q^2}{2} + 2q + C$

17. $\displaystyle\int 3\sin\theta\, d\theta = -3\cos\theta + C$

21. $\displaystyle\int (5\sin x + 3\cos x)\, dx = -5\cos x + 3\sin x + C$

25. $\displaystyle\int \pi r^2 h\, dr = \pi h\left(\dfrac{r^3}{3}\right) + C = \dfrac{\pi}{3}hr^3 + C$

29. $\displaystyle\int (3x^2 + 6e^{2x})\, dx = 3 \cdot \dfrac{x^3}{3} + 6 \cdot \dfrac{e^{2x}}{2} + C$
$$= x^3 + 3e^{2x} + C$$

33. We use the substitution $w = y + 2$, $dw = dy$:

$$\int \frac{1}{y + 2}\, dy = \int \frac{1}{w}\, dw = \ln|w| + C = \ln|y + 2| + C.$$

37. $\displaystyle\int \left(a\left(\dfrac{1}{x}\right) + bx^{-2}\right) dx = a\ln|x| + b\dfrac{x^{-1}}{-1} + C = a\ln|x| - \dfrac{b}{x} + C$

41. $\displaystyle\int P_0 e^{kt}\, dt = P_0\left(\dfrac{1}{k}e^{kt}\right) + C = \dfrac{P_0}{k}e^{kt} + C$

45. We use the substitution $w = 2 + e^x$, $dw = e^x\, dx$.

$$\int \frac{e^x}{2 + e^x}\, dx = \int \frac{dw}{w} = \ln|w| + C = \ln(2 + e^x) + C.$$

(We can drop the absolute value signs since $2 + e^x \geq 0$ for all x.)

Check: $\dfrac{d}{dx}[\ln(2 + e^x) + C] = \dfrac{1}{2 + e^x} \cdot e^x = \dfrac{e^x}{2 + e^x}.$

CHAPTER SEVEN

Solutions for Section 7.1

1. We use the fact that the area of a rectangle is Base $\times$ Height.
   (a) The fraction less than 5 meters high is the area to the left of 5, so
   $$\text{Fraction} = 5 \cdot 0.05 = 0.25.$$
   (b) The fraction above 6 meters high is the area to the right of 6, so
   $$\text{Fraction} = (20 - 6)0.05 = 0.7.$$
   (c) The fraction between 2 and 5 meters high is the area between 2 and 5, so
   $$\text{Fraction} = (5 - 2)0.05 = 0.15.$$

5. We can determine the fractions by estimating the area under the curve. Counting the squares for insects in the larval stage between 10 and 12 days we get about 4.5 squares, with each square representing $(2)\cdot(3\%)$ giving a total of 27% of the insects in the larval stage between 10 and 12 days.
   Likewise we get about 2 squares for the insects in the larval stage for less than 8 days, giving 12% of the insects in the larval stage for less than 8 days.
   Likewise we get about 7.5 squares for the insects in the larval stage for more than 12 days, giving 45% of the insects in the larval stage for more than 12 days.
   Since the peak of the graph occurs between 12 and 13 days, the length of the larval stage is most likely to fall in this interval.

9. Since $p(x)$ is a density function,
   $$\text{Area under graph} = \frac{1}{2} \cdot 50c = 25c = 1,$$
   so $c = 1/25 = 0.04$.

13. (a) The total area under the graph must be 1, so
    $$\text{Area } = 5(0.01) + 5C = 1$$
    So
    $$5C = 1 - 0.05 = 0.95$$
    $$C = 0.19$$
    (b) The machine is more likely to break in its tenth year than first year. It is equally likely to break in its first year and second year.
    (c) Since $p(t)$ is a density function,
    $$\text{Fraction of machines lasting up to 2 years} = \text{Area from 0 to 2} = 2(0.01) = 0.02.$$
    $$\text{Fraction of machines lasting between 5 and 7 years} = \text{Area from 5 to 7} = 2(0.19) = 0.38.$$
    $$\text{Fraction of machines lasting between 3 and 6 years} = \text{Area from 3 to 6}$$
    $$= \text{Area from 3 to 5} + \text{Area from 5 to 6}$$
    $$= 2(0.01) + 1(0.19) = 0.21.$$

17. See Figure 7.1. Many other answers are possible.

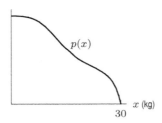

Figure 7.1

Solutions for Section 7.2

1. (a) The cumulative distribution function $P(t)$ is defined to be the fraction of patients who get in to see the doctor within t hours. No one gets in to see the doctor in less than 0 minutes, so $P(0) = 0$. We saw in Example 2 part (c) that 60% of patients wait less than 1 hour, so $P(1) = 0.60$. We saw in part (b) of Example 2 that an additional 37.5% of patients get in to see the doctor within the second hour, so 97.5% of patients will see the doctor within 2 hours; $P(2) = 0.975$. Finally, all patients are admitted within 3 hours, so $P(3) = 1$. Notice also that $P(t) = 1$ for all values of t greater than 3. A table of values for $P(t)$ is given in Table 7.1.

Table 7.1 *Cumulative distribution function for the density function in Example 2*

t	0	1	2	3	4	$\cdots$
$P(t)$	0	0.60	0.975	1	1	$\cdots$

(b) The graph is in Figure 7.2.

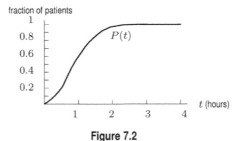

Figure 7.2

5. Since the function takes on the value of 4, it cannot be a cdf (whose maximum value is 1). In addition, the function decreases for $x > c$, which means that it is not a cdf. Thus, this function is a pdf. The area under a pdf is 1, so $4c = 1$ giving $c = \frac{1}{4}$. The pdf is $p(x) = 4$ for $0 \leq x \leq \frac{1}{4}$, so the cdf is given in Figure 7.3 by

$$P(x) = \begin{cases} 0 & \text{for} \quad x < 0 \\ 4x & \text{for} \quad 0 \leq x \leq \dfrac{1}{4} \\ 1 & \text{for} \quad x > \dfrac{1}{4} \end{cases}$$

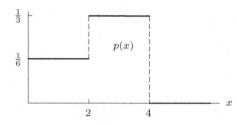

Figure 7.3

9. This function increases and levels off to c. The area under the curve is not finite, so it is not 1. Thus, the function must be a cdf, not a pdf, and $3c = 1$, so $c = 1/3$.

The pdf, $p(x)$ is the derivative, or slope, of the function shown, so, using $c = 1/3$,

$$p(x) = \begin{cases} 0 & \text{for} \quad x < 0 \\ (1/3 - 0)/(2 - 0) = 1/6 & \text{for} \quad 0 \le x \le 2 \\ (1 - 1/3)/(4 - 2) = 1/3 & \text{for} \quad 2 < x \le 4 \\ 0 & \text{for} \quad x > 4. \end{cases}$$

See Figure 7.4.

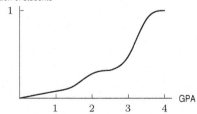

Figure 7.4

13. (a) The fraction of students passing is given by the area under the curve from 2 to 4 divided by the total area under the curve. This appears to be about $\frac{2}{3}$.
  (b) The fraction with honor grades corresponds to the area under the curve from 3 to 4 divided by the total area. This is about $\frac{1}{3}$.
  (c) The peak around 2 probably exists because many students work to get just a passing grade.
  (d)

fraction of students

17. (a) The cumulative distribution, $P(t)$, is the function whose slope is the density function $p(t)$. So $P'(t) = p(t)$. The graph of $P(t)$ starts out with a small slope at $t = 0$; its slope increases as t increases to 3. The graph of $P(t)$ levels off at 1 for $t \ge 4$. See Figure 7.5.

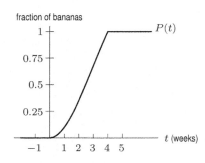

Figure 7.5

(b) The probability that a banana will last between 1 and 2 weeks is given by the difference $P(2) - P(1)$ where $P'(t) = p(t)$ and $p(t)$ is the density function. Looking at Figure 7.5 we see that the difference is roughly $0.25 = 25\%$.

21. (a) The probability you dropped the glove within a kilometer of home is given by

$$\int_0^1 2e^{-2x}\,dx = -e^{-2x}\Big|_0^1 = -e^{-2} + 1 \approx 0.865.$$

(b) Since the probability that the glove was dropped within y km $= \int_0^y p(x)dx = 1 - e^{-2y}$, we solve

$$1 - e^{-2y} = 0.95$$
$$e^{-2y} = 0.05$$
$$y = \frac{\ln 0.05}{-2} \approx 1.5 \text{ km.}$$

Solutions for Section 7.3

1. The median daily catch is the amount of fish such that half the time a boat will bring back more fish and half the time a boat will bring back less fish. Thus the area under the curve and to the left of the median must be 0.5. There are 25 squares under the curve so the median occurs at 12.5 squares of area. Now

$$\int_2^5 p(x)dx = 10.5 \text{ squares}$$

and

$$\int_5^6 p(x)dx = 5.5 \text{ squares,}$$

so the median occurs at a little over 5 tons. We must find the value a for which

$$\int_5^a p(t)dt = 2 \text{ squares,}$$

and we note that this occurs at about $a = 0.35$. Hence

$$\int_2^{5.35} p(t)\,dt \approx 12.5 \text{ squares}$$

$$\approx 0.5.$$

The median is about 5.35 tons.

5. We know that the median is given by T such that

$$\int_{-\infty}^{T} p(t)dt = 0.5.$$

Trying different values of T, we find that $0.5 = \int_0^T p(t)dt$ for $T \approx 2.48$ weeks. Figure 7.6 supports the conclusion that $t = 2.48$ is in fact the median.

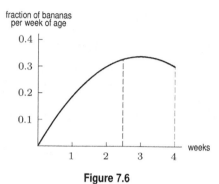

Figure 7.6

9. (a) The normal distribution of car speeds with $\mu = 58$ and $\sigma = 4$ is shown in Figure 7.7.

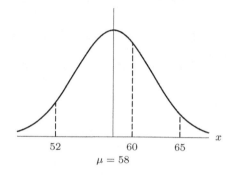

Figure 7.7

The probability that a randomly selected car is going between 60 and 65 is equal to the area under the curve from $x = 60$ to $x = 65$,

$$\text{Probability} = \frac{1}{4\sqrt{2\pi}} \int_{60}^{65} e^{-(x-58)^2/(2\cdot 4^2)} \, dx \approx 0.2685.$$

We obtain the value 0.2685 using a calculator or computer.

(b) To find the fraction of cars going under 52 km/hr, we evaluate the integral

$$\text{Fraction} = \frac{1}{4\sqrt{2\pi}} \int_{0}^{52} e^{-(x-58)^2/32} dx \approx 0.067.$$

Thus, approximately 6.7% of the cars are going less than 52 km/hr.

Solutions for Chapter 7 Review

1. Since $p(x)$ is a density function, the area under the graph of $p(x)$ is 1, so

$$\text{Area} = \text{Base} \cdot \text{Height} = 15a = 1$$

$$a = \frac{1}{15}.$$

5. Suppose x is the age of death; a possible density function is graphed in Figure 7.8.

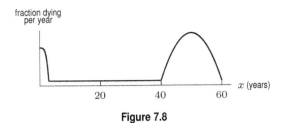

Figure 7.8

9.

Figure 7.9: Density function   **Figure 7.10**: Cumulative distribution function

13. The fact that most of the area under the graph of the density function is concentrated in two humps, centered at 8 and 12 years, indicates that most of the population belong to one of two groups, those who leave school after finishing approximately 8 years and those who finish about 12 years. There is a smaller group of people who finish approximately 16 years of school.

  The percentage of adults who have completed less than ten years of school is equal to the area under the density function to the left of the vertical line at $t = 10$. (See Figure 7.11.) We know that the total area is 1, so we are estimating the percentage of the total area that lies in this shaded part shown in Figure 7.11. A rough estimate of this area is about 30%.

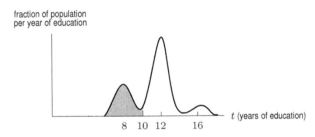

Figure 7.11: What percent has less than 10 years of education?

17. Figure 7.12 is a graph of the density function; Figure 7.13 is a graph of the cumulative distribution.

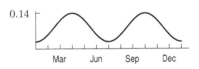

Figure 7.12: Density function

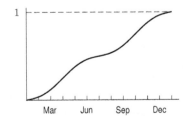

Figure 7.13: Cumulative distribution function

21. False. Note that p is the density function for the population, not the cumulative density function. Thus $p(10) = 1/2$ means that the probability of x lying in a small interval of length Δx around $x = 10$ is about $(1/2)\Delta x$.

25. True. By the definition of the cumulative distribution function, $P(20) - P(10) = 0$ is the fraction of the population having x values between 10 and 20.

STRENGTHEN YOUR UNDERSTANDING

1. False, since we also need $p(x) \geq 0$.

5. False. The integral $\int_0^{70} p(x)\, dx$ represents the fraction of the population between the ages of 0 and 70 years old. The fraction will be significantly more than 50% of the population.

9. True, since the population with values between 0 and 1 is included in the population with values between 0 and 2.

13. False. $P(30)$ is the fraction of the population with values at or below 30.

17. False. The units of $p(x)$ are fraction of population per unit of x, while the units for $P(x)$ are fraction of population.

21. True, this is the definition of mean value as given in the text.

25. True, since the median T is the value for which one-half of the population is greater than or equal to T.

29. True, since the integrand is a normal distribution with mean $\mu = 7$ and standard deviation $\sigma = 1$.

CHAPTER EIGHT

Solutions for Section 8.1

1. We make a table by calculating values for $C = f(d, m)$ for each value of d and m. Such a table is shown in Table 8.1

Table 8.1

		d			
		1	2	3	4
	100	55	95	135	175
m	200	70	110	150	190
	300	85	125	165	205
	400	100	140	180	220

5. If the price of beef is held constant, beef consumption for households with various incomes can be read from a fixed column in Table 8.2 on page 356 of the text. For example, the column corresponding to $p = 3.00$ gives the function $h(I) = f(I, 3.00)$; it tells you how much beef a household with income I will buy at \$3.00/lb. Looking at the column from the top down, you can see that it is an increasing function of I. This is true in every column. This says that at any fixed price for beef, consumption goes up as household income goes up—which makes sense. Thus, f is an increasing function of I for each value of p.

9. Asking if f is an increasing or decreasing function of p is the same as asking how does f vary as we vary p, when we hold a fixed. Intuitively, we know that as we increase the price p, total sales of the product will go down. Thus, f is a decreasing function of p. Similarly, if we increase a, the amount spent on advertising, we can expect f to increase and therefore f is an increasing function of a.

13. We expect P to be an increasing function of A and r. (If you borrow more, your payments go up; if the interest rates go up, your payments go up.) However, P is a decreasing function of t. (If you spread out your payments over more years, you pay less each month.)

Solutions for Section 8.2

1. We can see that as we move horizontally to the right, we are increasing x but not changing y. As we take such a path at $y = 2$, we cross decreasing contour lines, starting at the contour line 6 at $x = 1$ to the contour line 1 at around $x = 5.7$. This trend holds true for all of horizontal paths. Thus, z is a decreasing function of x. Similarly, as we move up along a vertical line, we cross increasing contour lines and thus z is an increasing function of y.

5. The contour where $f(x, y) = x + y = c$, or $y = -x + c$, is the graph of the straight line with slope -1 as shown in Figure 8.1. Note that we have plotted the contours for $c = -3, -2, -1, 0, 1, 2, 3$. The contours are evenly spaced.

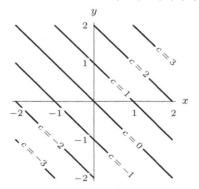

Figure 8.1

9. The contour where $f(x, y) = -x - y = c$ or $y = -x - c$ is the graph of the straight line of slope -1 as shown in Figure 8.2. Note that we have plotted contours for $c = -3, -2, -1, 0, 1, 2, 3$. The contours are evenly spaced.

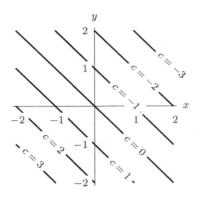

Figure 8.2

13. The contour for $C = 50$ is given by

$$40d + 0.15m = 50.$$

This is the equation of a line with intercepts $d = 50/40 = 1.25$ and $m = 50/0.15 \approx 333$. (See Figure 8.3.) The contour for $C = 100$ is given by

$$40d + 0.15m = 100.$$

This is the equation of a parallel line with intercepts $d = 100/40 = 2.5$ and $m = 100/0.15 \approx 667$. The contours for $C = 150$ and $C = 200$ are parallel lines drawn similarly. (See Figure 8.3.)

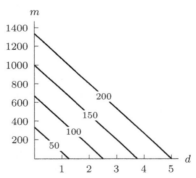

Figure 8.3: A contour diagram for
$C = 40d + 0.15m$

17.

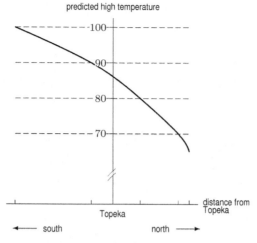

Figure 8.4

21. The contours of $z = y - \sin x$ are of the form $y = \sin x + c$ for a constant c. They are sinusoidal graphs shifted vertically by the value of z on the contour line. The contours are equally spaced vertically for equally spaced z values. See Figure 8.5.

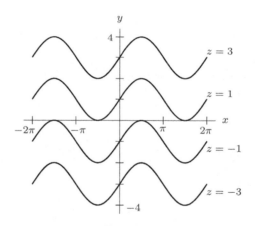

Figure 8.5

25. (a) Find the point where the horizontal line for 15 mph meets the contour for $-20°$F wind chill. The actual temperature is about $0°$F.
  (b) The horizontal line for 10 mph meets the vertical line for $0°$F about $1/5$ of the way from the contour for $-20°$F to the contour for $0°$F wind chill. We estimate the wind chill to be about $-16°$F.
  (c) We look for the point on the vertical line for $-20°$F where the wind chill is $-50°$F, the danger point for humans. This is a point on the line that is about half way between the contours for $-60°$F and $-40°$F. The point can not be determined exactly, but we estimate that it occurs where the wind speed is about 23 mph.
  (d) A temperature drop of $20°$F corresponds to moving left from one vertical grid line to the next on the horizontal line for 15 mph. This horizontal movement appears to correspond to about 1 1/4 the horizontal distance between contours crossing the line. Since contours are spaced at $20°$F wind chill, we estimate that the wind chill drops about $25°$F when the air temperature goes down $20°$F during a 15 mph wind.

29. (a) The TMS map of an eye of constant curvature will have only one color, with no contour lines dividing the map.
  (b) The contour lines are circles, because the cross-section is the same in every direction. The largest curvature is in the center. See picture below.

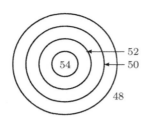

33. **(a)** is (II). According to the diagram, increasing X without changing Y leads to greater satisfaction, but increasing Y without changing X leads to reduced satisfaction. More X is desirable, and more Y is undesirable. This fits the option where X is income and Y is hours worked.
   (b) is (I). According to the diagram, increasing X without changing Y and increasing Y without changing X both lead to greater satisfaction. This fits the option where X is income and Y is leisure time.

Solutions for Section 8.3

1. **(a)** Positive.
   (b) Negative.
   (c) Positive.
   (d) Zero.

5. **(a)** We expect the demand for coffee to decrease as the price of coffee increases (assuming the price of tea is fixed.) Thus we expect f_c to be negative. We expect people to switch to coffee as the price of tea increases (assuming the price of coffee is fixed), so that the demand for coffee will increase. We expect f_t to be positive.
   (b) The statement $f(3,2) = 780$ tells us that if coffee costs \$3 per pound and tea costs \$2 per pound, we can expect 780 pounds of coffee to sell each week. The statement $f_c(3,2) = -60$ tells us that, if the price of coffee then goes up \$1 and the price of tea stays the same, the demand for coffee will go down by about 60 pounds. The statement $20 = f_t(3,2)$ tells us that if the price of tea goes up \$1 and the price of coffee stays the same, the demand for coffee will go up by about 20 pounds.

9. We know $z_x(1,0)$ is the rate of change of z in the x-direction at $(1,0)$. Therefore

$$z_x(1,0) \approx \frac{\Delta z}{\Delta x} \approx \frac{1}{0.5} = 2, \quad \text{so } z_x(1,0) \approx 2.$$

We know $z_x(0,1)$ is the rate of change of z in the x-direction at the point $(0,1)$. Since we move along the contour, the change in z

$$z_x(0,1) \approx \frac{\Delta z}{\Delta x} \approx \frac{0}{\Delta x} = 0.$$

We know $z_y(0,1)$ is the rate of change of z in the y-direction at the point $(0,1)$ so

$$z_y(0,1) \approx \frac{\Delta z}{\Delta y} \approx \frac{1}{0.1} = 10.$$

13. Estimate $\partial P/\partial r$ and $\partial P/\partial L$ by using difference quotients and reading values of P from the graph:

$$\left.\frac{\partial P}{\partial r}\right|_{(8,5000)} \approx \frac{P(15,5000) - P(8,5000)}{15 - 8}$$
$$= \frac{120 - 100}{7} = 2.9,$$

and

$$\left.\frac{\partial P}{\partial L}\right|_{(8,5000)} \approx \frac{P(8,4000) - P(8,5000)}{4000 - 5000}$$
$$= \frac{80 - 100}{-1000} = 0.02.$$

$P_r(8, 5000) \approx 2.9$ means that at an interest rate of 8% and a loan amount of $5000 the monthly payment increases by approximately $2.90 for an additional one percent increase of the interest rate. $P_L(8, 5000) \approx 0.02$ means the monthly payment increases by approximately $0.02 for an additional $1 increase in the loan amount at an 8% rate and a loan amount of $5000.

17. (a) The table gives $f(200, 400) = 150,000$. This means that sales of 200 full-price tickets and 400 discount tickets generate $150,000 in revenue.

(b) The notation $f_x(200, 400)$ represents the rate of change of f as we fix y at 400 and increase x from 200. What happens to the revenue as we look along the row $y = 400$ in the table? Revenue increases, so $f_x(200, 400)$ is positive. The notation $f_y(200, 400)$ represents the rate of change of f as we fix x at 200 and increase y from 400. What happens to the revenue as we look down the column $x = 200$ in the table? Revenue increases, so $f_y(200, 400)$ is positive.

Both partial derivatives are positive. This makes sense since revenue increases if more of either type of ticket is sold.

(c) To estimate $f_x(200, 400)$, we calculate $\Delta R/\Delta x$ as x increases from 200 to 300 while y is fixed at 400. We have

$$f_x(200, 400) \approx \frac{\Delta R}{\Delta x} = \frac{185,000 - 150,000}{300 - 200} = 350 \text{ dollars/ticket.}$$

The partial derivative of f with respect to x is 350 dollars per full-price ticket. This means that the price of a full-price ticket is $350.

To estimate $f_y(200, 400)$, we calculate $\Delta R/\Delta y$ as y increases from 400 and 600 while x is fixed at 200. We have

$$f_y(200, 400) \approx \frac{\Delta R}{\Delta y} = \frac{190,000 - 150,000}{600 - 400} = 200 \text{ dollars/ticket.}$$

The partial derivative of f with respect to y is 200 dollars per discount ticket. This means that the price of a discount ticket is $200.

21. We are asked to find $f(26, 15)$. The closest to this point in the data given in the table is $f(24, 15)$, so we will approximate based on this value. First we need to calculate the partial derivative f_t. Using $\Delta t = -2$, we have

$$f(24, 15) \approx \frac{f(22, 15) - f(24, 15)}{-2} = \frac{58 - 36}{-2} = -11.$$

Given this, we can use $\Delta f \approx \Delta t f_t + \Delta c f_c$ to obtain

$$f(26, 15) \approx f(24, 15) + 2 \cdot f_t(24, 15) + 0 \cdot f_c(26, 15)$$
$$\approx 36 - 2 \cdot 11 + 0$$
$$= 14\%.$$

25. (a) Since the contours are parallel straight lines, they all have the same slope. Two points on the contour for happiness level 10 are (0 cherries, 10 grapes) and (5 cherries, 0 grapes). Thus

$$\text{Slope} = \frac{10 - 0}{0 - 5} = -2 \frac{\text{Grapes}}{\text{Cherry}}.$$

(b) Dan's happiness with his snack stays the same if he replaces two grapes with one cherry.

Solutions for Section 8.4

1. $f_x(x, y) = 2x + 2y$, $f_y(x, y) = 2x + 3y^2$.

5. $\dfrac{\partial z}{\partial x} = 2xe^y$

9. $f_x(x, y) = 20xe^{3y}$, $f_y(x, y) = 30x^2 e^{3y}$.

13. $\dfrac{\partial A}{\partial h} = \dfrac{1}{2}(a + b)$

17. **(a)** From contour diagram,

$$f_x(2,1) \approx \frac{f(2.3,1) - f(2,1)}{2.3 - 2}$$

$$= \frac{6-5}{0.3} = 3.3,$$

$$f_y(2,1) \approx \frac{f(2,1.4) - f(2,1)}{1.4 - 1}$$

$$= \frac{6-5}{0.4} = 2.5.$$

(b) A table of values for f is given in Table 8.2.

Table 8.2

		\multicolumn{3}{c}{y}	
		0.9	1.0	1.1
	1.9	4.42	4.61	4.82
x	2.0	4.81	5.00	5.21
	2.1	5.22	5.41	5.62

From Table 8.2 we estimate $f_x(2,1)$ and $f_y(2,1)$ using difference quotients:

$$f_x(2,1) \approx \frac{5.41 - 5.00}{2.1 - 2} = 4.1$$

$$f_y(2,1) \approx \frac{5.21 - 5.00}{1.1 - 1} = 2.1.$$

(c) $f_x(x,y) = 2x$, $f_y(x,y) = 2y$. So the exact values are $f_x(2,1) = 4$, $f_y(2,1) = 2$.

21. $f_x = 2xy$ and $f_y = x^2$, so $f_{xx} = 2y$, $f_{xy} = 2x$, $f_{yy} = 0$ and $f_{yx} = 2x$.

25. $f_x = 2xy^2$ and $f_y = 2x^2y$, so $f_{xx} = 2y^2$, $f_{xy} = 4xy$, $f_{yy} = 2x^2$ and $f_{yx} = 4xy$.

29. $f_r = 100te^{rt}$ and $f_t = 100re^{rt}$, so $f_{rr} = 100t^2e^{rt}$, $f_{rt} = f_{tr} = 100tre^{rt} + 100e^{rt} = 100(rt + 1)e^{rt}$ and $f_{tt} = 100r^2e^{rt}$.

33. Since $f_x(x,y) = 4x^3y^2 - 3y^4$, we could have

$$f(x,y) = x^4y^2 - 3xy^4.$$

In that case,

$$f_y(x,y) = \frac{\partial}{\partial y}(x^4y^2 - 3xy^4) = 2x^4y - 12xy^3$$

as expected. More generally, we could have $f(x,y) = x^4y^2 - 3xy^4 + C$, where C is any constant.

37. **(a)** The quotient rule gives

$$\frac{\partial M}{\partial c} = \frac{(c+r)1 - (c+1)1}{(c+r)^2}B = -\frac{1-r}{(c+r)^2}B.$$

(b) Negative sign, because $c > 0$, $0 < r < 1$, and $B > 0$.

(c) Suppose that banks hold a constant ratio of cash to checking account balances, and that the amount of cash in an economy is constant. If people start to keep a greater fraction of their money in cash and less in checking accounts, then the money supply decreases.

Solutions for Section 8.5

1. We can identify local extreme points on a contour diagram because these points will either be the centers of a series of concentric circles that close around them, or will lie on the edges of the diagram. Looking at the graph, we see that $(2,10)$, $(6,4)$, $(6.5,16)$ and $(9,10)$ appear to be such points. Since the points near $(2,10)$ decrease in functional value

as they close around $(2, 10)$, $f(2, 10)$ will be somewhat less than its nearest contour. So $f(2, 10) \approx 0.5$. Similarly, since the contours near $(2, 10)$ are greater in functional value than $f(2, 10)$, $f(2, 10)$ is a local minimum. Applying analogous arguments to the point $(6, 4)$, we see that $f(6, 4) \approx 9.5$ and is a local maximum. The contour values are increasing as we approach $(6.5, 16)$ along any path, so $f(6.5, 16) \approx 10$ is a local maximum and $(9, 10)$ is a local minimum.

Since none of the local minima are less in value than $f(2, 10) \approx 0.5$, $f(2, 10)$ is a global minimum. Since none of the local maxima are greater in value than $f(6.5, 16) \approx 10$, $f(6.5, 16)$ is a global maximum.

5. The maxima occur at about $(\pi/2, 0)$ and $(\pi/2, 2\pi)$. The minimum occurs at $(\pi/2, \pi)$. The maximum value is about 1, the minimum value is about -1.

9. At a critical point $f_x = -3y + 6 = 0$ and $f_y = 3y^2 - 3x = 0$, so $(4, 2)$ is the only critical point. Since $f_{xx}f_{yy} - f_{xy}^2 = -9 < 0$, the point $(4, 2)$ is a saddle point.

13. To find the critical points, we solve $f_x = 0$ and $f_y = 0$ for x and y. Solving

$$f_x = 2x - 2y = 0,$$
$$f_y = -2x + 6y - 8 = 0.$$

We see from the first equation that $x = y$. Substituting this into the second equation shows that $y = 2$. The only critical point is $(2, 2)$.

We have

$$D = (f_{xx})(f_{yy}) - (f_{xy})^2 = (2)(6) - (-2)^2 = 8.$$

Since $D > 0$ and $f_{xx} = 2 > 0$, the function f has a local minimum at the point $(2, 2)$.

17. At a local maximum value of f,

$$\frac{\partial f}{\partial x} = -2x - B = 0.$$

We are told that this is satisfied by $x = -2$. So $-2(-2) - B = 0$ and $B = 4$. In addition,

$$\frac{\partial f}{\partial y} = -2y - C = 0$$

and we know this holds for $y = 1$, so $-2(1) - C = 0$, giving $C = -2$. We are also told that the value of f is 15 at the point $(-2, 1)$, so

$$15 = f(-2, 1) = A - ((-2)^2 + 4(-2) + 1^2 - 2(1)) = A - (-5), \text{ so } A = 10.$$

Now we check that these values of A, B, and C give $f(x, y)$ a local maximum at the point $(-2, 1)$. Since

$$f_{xx}(-2, 1) = -2,$$
$$f_{yy}(-2, 1) = -2$$

and

$$f_{xy}(-2, 1) = 0,$$

we have that $f_{xx}(-2, 1)f_{yy}(-2, 1) - f_{xy}^2(-2, 1) = (-2)(-2) - 0 > 0$ and $f_{xx}(-2, 1) < 0$. Thus, f has a local maximum value 15 at $(-2, 1)$.

21. The total revenue is

$$R = pq = (60 - 0.04q)q = 60q - 0.04q^2,$$

and as $q = q_1 + q_2$, this gives

$$R = 60q_1 + 60q_2 - 0.04q_1^2 - 0.08q_1q_2 - 0.04q_2^2.$$

Therefore, the profit is

$$P(q_1, q_2) = R - C_1 - C_2$$
$$= -13.7 + 60q_1 + 60q_2 - 0.07q_1^2 - 0.08q_2^2 - 0.08q_1q_2.$$

At a local maximum point, we would have:

$$\frac{\partial P}{\partial q_1} = 60 - 0.14q_1 - 0.08q_2 = 0,$$
$$\frac{\partial P}{\partial q_2} = 60 - 0.16q_2 - 0.08q_1 = 0.$$

Solving these equations, we find that

$$q_1 = 300 \quad \text{and} \quad q_2 = 225.$$

To see whether or not we have found a local maximum, we compute the second-order partial derivatives:

$$\frac{\partial^2 P}{\partial q_1^2} = -0.14, \quad \frac{\partial^2 P}{\partial q_2^2} = -0.16, \quad \frac{\partial^2 P}{\partial q_1 \partial q_2} = -0.08.$$

Therefore,

$$D = \frac{\partial^2 P}{\partial q_1^2} \frac{\partial^2 P}{\partial q_2^2} - \frac{\partial^2 P}{\partial q_1 \partial q_2} = (-0.14)(-0.16) - (-0.08)^2 = 0.016,$$

and so we have found a local maximum point. The graph of $P(q_1, q_2)$ has the shape of an upside down paraboloid since P is quadratic in q_1 and q_2, hence $(300, 225)$ is a global maximum point.

Solutions for Section 8.6

1. Our objective function is $f(x, y) = x + y$ and our equation of constraint is $g(x, y) = x^2 + y^2 = 1$. To optimize $f(x, y)$ with Lagrange multipliers, we solve the following system of equations

$$f_x(x, y) = \lambda g_x(x, y), \quad \text{so } 1 = 2\lambda x$$
$$f_y(x, y) = \lambda g_y(x, y), \quad \text{so } 1 = 2\lambda y$$
$$g(x, y) = 1, \quad \text{so } x^2 + y^2 = 1$$

Solving for λ gives

$$\lambda = \frac{1}{2x} = \frac{1}{2y},$$

which tells us that $x = y$. Going back to our equation of constraint, we use the substitution $x = y$ to solve for y:

$$g(y, y) = y^2 + y^2 = 1$$
$$2y^2 = 1$$
$$y^2 = \frac{1}{2}$$
$$y = \pm\sqrt{\frac{1}{2}} = \pm\frac{\sqrt{2}}{2}.$$

Since $x = y$, our critical points are $(\frac{\sqrt{2}}{2}, \frac{\sqrt{2}}{2})$ and $(-\frac{\sqrt{2}}{2}, -\frac{\sqrt{2}}{2})$. Since the constraint is closed and bounded, maximum and minimum values of f subject to the constraint exist. Evaluating f at the critical points we find that the maximum value is $f(\frac{\sqrt{2}}{2}, \frac{\sqrt{2}}{2}) = \sqrt{2}$ and the minimum value is $f(-\frac{\sqrt{2}}{2}, -\frac{\sqrt{2}}{2}) = -\sqrt{2}$.

5. We wish to optimize $f(x, y) = 5xy$ subject to the constraint $g(x, y) = x + 3y = 24$. To do this we must solve the following system of equations:

$$f_x(x, y) = \lambda g_x(x, y), \text{ so } 5y = \lambda$$
$$f_y(x, y) = \lambda g_y(x, y), \text{ so } 5x = 3\lambda$$
$$g(x, y) = 24, \text{ so } x + 3y = 24$$

Solving these equations produces:

$$x = 12 \quad y = 4 \quad \lambda = 20$$

corresponding to optimal $f(x, y) = 5(12)(4) = 240$.

9. The objective function is $f(x, y) = x^2 + y^2$ and the constraint equation is $g(x, y) = 4x - 2y = 15$, so $f_x = 2x, f_y = 2y$ and $g_x = 4, g_y = -2$. We have:

$$2x = 4\lambda,$$
$$2y = -2\lambda.$$

From the first equation we have $\lambda = x/2$, and from the second equation we have $\lambda = -y$. Setting these equal gives

$$y = -0.5x.$$

Substituting this into the constraint equation $4x - 2y = 15$ gives $x = 3$. The only critical point is $(3, -1.5)$.

We have $f(3, -1.5) = (3)^2 + (1.5)^2 = 11.25$. One way to determine if this point gives a maximum or minimum value or neither for the given constraint is to examine the contour diagram of f with the constraint sketched in, Figure 8.6. It appears that moving away from the point $P = (3, -1.5)$ in either direction along the constraint increases the value of f, so $(3, -1.5)$ is a point of minimum value.

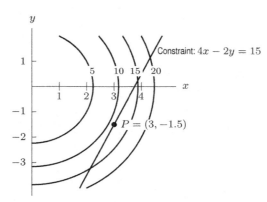

Figure 8.6

13. (a) Objective function: $Q = x_1^{0.3} x_2^{0.7}$.
    (b) Constraint: $10x_1 + 25x_2 = 50,000$.

17. (a) To be producing the maximum quantity Q under the cost constraint given, the firm should be using K and L values given by

$$\frac{\partial Q}{\partial K} = 0.6aK^{-0.4}L^{0.4} = 20\lambda$$

$$\frac{\partial Q}{\partial L} = 0.4aK^{0.6}L^{-0.6} = 10\lambda$$

$$20K + 10L = 150.$$

Hence $\dfrac{0.6aK^{-0.4}L^{0.4}}{0.4aK^{0.6}L^{-0.6}} = 1.5\dfrac{L}{K} = \dfrac{20\lambda}{10\lambda} = 2$, so $L = \dfrac{4}{3}K$. Substituting in $20K + 10L = 150$, we obtain $20K + 10\left(\dfrac{4}{3}\right)K = 150$. Then $K = \dfrac{9}{2}$ and $L = 6$, so capital should be reduced by $\dfrac{1}{2}$ unit, and labor should be increased by 1 unit.

    (b) $\dfrac{\text{New production}}{\text{Old production}} = \dfrac{a4.5^{0.6}6^{0.4}}{a5^{0.6}5^{0.4}} \approx 1.01$, so tell the board of directors, "Reducing the quantity of capital by 1/2 unit and increasing the quantity of labor by 1 unit will increase production by 1% while holding costs to \$150."

21. (a) The objective function is the function that is optimized. Since the problem refers to maximum production, the objective function is the production function $P(K, L)$.
    (b) Production is maximized subject to a budget restriction, which is the constraint. The constraint equation is $C(K, L) = 600,000$.
    (c) The Lagrange multiplier tells you the rate at which maximum production changes when the budget is increased. Its units are tons of steel per dollar of budget, or simply tons/dollar.
    (d) Increasing the budget from \$600,000 to \$(600,000+a) increases the maximum possible production from 2,500,000 tons to approximately $(2,500,000 + 3.17a)$ tons. Every extra dollar of budget increases maximal production by approximately $\lambda = 3.17$ tons.

25. The constraint is

$$g(x_1, x_2) = x_1 + 3x_2 = 100.$$

We need to solve the equations

$$\frac{\partial U}{\partial x_1} = \lambda \frac{\partial g}{\partial x_1}, \qquad \frac{\partial U}{\partial x_2} = \lambda \frac{\partial g}{\partial x_2}, \qquad x_1 + 3x_2 = 100.$$

These equations are

$$2x_2 + 3 = \lambda$$
$$2x_1 = 3\lambda$$
$$x_1 + 3x_2 = 100.$$

The first equation gives

$$x_2 = \frac{\lambda - 3}{2}$$

and the second that

$$x_1 = \frac{3\lambda}{2}.$$

Substituting these into the third equation gives

$$\frac{3\lambda}{2} + 3\left(\frac{\lambda - 3}{2}\right) = 100,$$

therefore, $\lambda = 209/6$ so $x_1 = 209/4$ and $x_2 = 191/12$. The maximum utility is

$$U\left(\frac{209}{4}, \frac{191}{12}\right) = 2\frac{209}{4}\frac{191}{12} + 3\frac{209}{4} = 1820.04.$$

The approximate change in maximum utility due to a one unit increase in the consumer's disposable income is λ. For a $6 increase in disposable income, the maximum utility increases by about 6λ which is 209.

Solutions for Chapter 8 Review

1. To see whether f is an increasing or decreasing function of x, we need to see how f varies as we increase x and hold y fixed. We note that each column of the table corresponds to a fixed value of y. Scanning down the $y = 2$ column, we can see that as x increases, the value of the function decreases from 114 when $x = 0$ down to 93 when $x = 80$. Thus, f may be decreasing. In order for f to actually be decreasing however, we have to make sure that f decreases for *every* column. In this case, we see that f indeed does decrease for every column. Thus, f is a decreasing function of x. Similarly, to see whether f is a decreasing function of y we need to look at the rows of the table. As we can see, f increases for every row as we increase y. Thus, f is an increasing function of y.

5. To draw a contour for a wind-chill of $W = 20$, we need a few combinations of temperature and wind velocity (T, v) such that $W(T, v) = 20$. Estimating from the table, some such points are $(24, 5)$ and $(33, 10)$. We can connect these points to get a contour for $W = 20$. Similarly, some points that have wind-chill of about $0° F$ are $(5, 5)$, $(17.5, 10)$, $(23.5, 15)$, $(27, 20)$, and $(29, 25)$. By connecting these points we get the contour for $W = 0$. If we carry out this procedure for more values of W, we get a full contour diagram such as is shown in Figure 8.7:

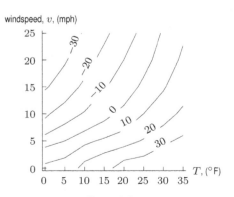

Figure 8.7

9. Looking at the contour diagram, we can see that $Q(x, y)$ is an increasing function of x and a decreasing function of y. It stands to reason that as the price of orange juice goes up, the demand for orange juice will go down. Thus, the demand is a decreasing function of the price of orange juice and thus the y-axis corresponds to the price of orange juice. Also, as the price of apple juice goes up, so will the demand for orange juice and therefore the demand for orange juice is an increasing function of the price of apple juice. Thus, the price of apple juice corresponds to the x-axis.

13. (a) The point representing 8% and $6000 on the graph lies between the 120 and 140 contours. We estimate the monthly payment to be about $122.
(b) Since the interest rate has dropped, we will be able to borrow more money and still make a monthly payment of $122. To find out how much we can afford to borrow, we find where the interest rate of 6% intersects the $122 contour and read off the loan amount to which these values correspond. Since the $122 contour is not shown, we estimate its position from the $120 and $140 contours. We find that we can borrow an amount of money that is more than $6000 but less than $6500. So we can borrow about $350 more without increasing the monthly payment.
(c) The entries in the table will be the amount of loan at which each interest rate intersects the 122 contour. Using the $122 contour from (b) we make table 8.3.

Table 8.3 *Amount borrowed at a monthly payment of $122.*

Interest Rate (%)	0	1	2	3	4	5	6	7
Loan Amount ($)	7400	7200	7000	6800	6650	6500	6350	6200
Interest rate (%)	8	9	10	11	12	13	14	15
Loan Amount ($)	6000	5850	5700	5600	5500	5400	5300	5200

17. (a) If you borrow $8000 at an interest rate of 1% per month and pay it off in 24 months, your monthly payments are $376.59.
(b) The increase in your monthly payments for borrowing an extra dollar under the same terms as in (a) is about 4.7 cents.
(c) If you borrow the same amount of money for the same time period as in (a), but if the interest rate increases by 1%, the increase in your monthly payments is about $44.83.

21. $\dfrac{\partial q}{\partial I} > 0$ because, other things being constant, as people get richer, more beer will be bought.

$\dfrac{\partial q}{\partial p_1} < 0$ because, other things being constant, if the price of beer rises, less beer will be bought.

$\dfrac{\partial q}{\partial p_2} > 0$ because, other things being constant, if the price of other goods rises, but the price of beer does not, more beer will be bought.

25. $f_x = 2x + y,\ f_y = 2y + x.$

29. $\dfrac{\partial P}{\partial K} = 7K^{-0.3}L^{0.3},\ \dfrac{\partial P}{\partial L} = 3K^{0.7}L^{-0.7}.$

33. (a)

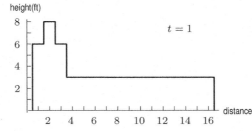

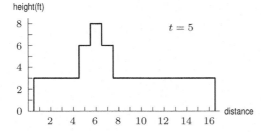
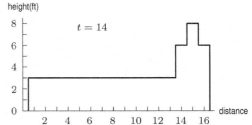

(b)

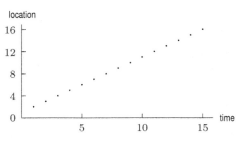

Figure 8.8

(c) The "wave" at a sports arena.

37. (a) We first express the revenue R in terms of the prices p_1 and p_2:

$$R(p_1, p_2) = p_1 q_1 + p_2 q_2$$
$$= p_1(517 - 3.5p_1 + 0.8p_2) + p_2(770 - 4.4p_2 + 1.4p_1)$$
$$= 517p_1 - 3.5p_1^2 + 770p_2 - 4.4p_2^2 + 2.2p_1 p_2.$$

(b) We compute the partial derivatives and set them to zero:

$$\frac{\partial R}{\partial p_1} = 517 - 7p_1 + 2.2p_2 = 0,$$

$$\frac{\partial R}{\partial p_2} = 770 - 8.8p_2 + 2.2p_1 = 0.$$

Solving these equations, we find that

$$p_1 = 110 \quad \text{and} \quad p_2 = 115.$$

To see whether or not we have a found a local maximum, we compute the second-order partial derivatives:

$$\frac{\partial^2 R}{\partial p_1^2} = -7, \quad \frac{\partial^2 R}{\partial p_2^2} = -8.8, \quad \frac{\partial^2 R}{\partial p_1 \partial p_2} = 2.2.$$

Therefore,

$$D = \frac{\partial^2 R}{\partial p_1^2}\frac{\partial^2 R}{\partial p_2^2} - \left(\frac{\partial^2 R}{\partial p_1 \partial p_2}\right)^2 = (-7)(-8.8) - (2.2)^2 = 56.76,$$

and so we have found a local maximum point. The graph of $R(p_1, p_2)$ has the shape of an upside down bowl. Therefore, $(110, 115)$ is a global maximum point.

41. (a) Objective function: $C = 127x_1 + 92x_2$.
   (b) Constraint: $x_1^{0.6} x_2^{0.4} = 500$.
   (c) The value of λ tells us that if the constraint is relaxed by one unit, the objective function changes by about λ units. In this context, this means that if the production quota of 500 is relaxed to 499, the cost decreases by about \$219.

45. (a) The profit is given by

$$\text{Profit} = \pi(q_1, q_2)$$
$$= \text{Total Revenue} - \text{Total Cost}$$
$$= p_1 q_1 + p_2 q_2 - (10q_1 + q_1 q_2 + 10q_2)$$
$$= (50 - q_1 + q_2)q_1 + (30 + 2q_1 - q_2)q_2 - (10q_1 + q_1 q_2 + 10q_2)$$
$$= 40q_1 - q_1^2 + 2q_1 q_2 + 20q_2 - q_2^2,$$

subject to the constraint $q_1 + q_2 = 15$.
   Then we need to solve the equations

$$\frac{\partial \pi}{\partial q_1} = 0, \quad \frac{\partial \pi}{\partial q_2} = 0, \text{ subject to } q_1 + q_2 = 15.$$

These equations are

$$40 - 2q_1 + 2q_2 = \lambda$$
$$2q_1 + 20 - 2q_2 = \lambda$$
$$q_1 + q_2 = 15.$$

Adding the first two equations gives

$$60 = 2\lambda$$

so $\lambda = 30$. Substituting this into the first equation gives

$$q_1 - q_2 = 5,$$

therefore, q_1 and q_2 satisfy

$$q_1 + q_2 = 15$$
$$q_1 - q_2 = 5.$$

Adding these equations gives $2q_1 = 20$ so $q_1 = 10$ and $q_2 = 5$. Substituting these values into the expression for the total profit gives

$$\pi(10, 5) = 40 \cdot 10 - 10^2 + 2 \cdot 10 \cdot 5 + 20 \cdot 5 - 5^2 = 475.$$

The endpoints of the constraint are $(15, 0)$ and $(0, 15)$ giving

$$\pi(15, 0) = 40 \cdot 15 - 15^2 = 375$$
$$\pi(0, 15) = 20 \cdot 15 - 15^2 = 75.$$

Thus the maximum profit is 475.

(b) The approximate change in the maximum profit due to a one unit increase in the production constraint is $\lambda = 30$. Thus a one unit increase in the production quota increases production by 30 units, to $475 + 30 = 505$ units.

STRENGTHEN YOUR UNDERSTANDING

1. False, the units of 1.5 are kilometers.

5. True.

9. False. The cross-section with $x = 1$ is $f(1, y) = e^y - y^2$.

13. True. If they intersected at some point (a, b), we would have simultaneously $f(a, b) = 1$ and $f(a, b) = 2$. This is impossible since a function can have only one output for a given input.

17. True, since a contour is defined as the set of (x, y) such that $f(x, y)$ is constant.

21. True, as specified in the text.

25. False. In general, as calorie consumption goes up, the weight goes up, so the partial derivative would be positive.

29. True, since when y is held constant at 3, we have $f_x(5, 3) \approx \Delta z / \Delta x = 0.9/0.1 = 9$.

33. True, since $g_u = e^v$ so $g_u(0, 0) = e^0 = 1$.

37. False. For example, let $f(x, y) = x^2 + y$. Then $f_{xx} = 2$ but $f_{yy} = 0$. It is true that the *mixed* second order partial derivatives are equal: $f_{xy} = f_{yx}$ if f_{xx}, f_{xy}, f_{yx}, and f_{yy} are all continuous.

41. False. We need also $f_y(1, 2) = 0$.

45. False. For example, let $f(x, y) = x^2 - y^2$. Then critical point $(0, 0)$ is neither a local maximum nor a local minimum.

49. True. We have $D = (0)(0) - 3^2 < 0$, so the critical point $(0, 0)$ is a saddle point.

53. False. The budget equation is the constraint equation.

57. True, since in this situation, λ represents the approximate change in cost given a one unit increase in production.

Solutions to Problems on Deriving the Formula for Regression Lines

1. Let the line be in the form $y = b + mx$. When x equals $-1, 0$ and 1, then y equals $b - m$, b, and $b + m$, respectively. The sum of the squares of the vertical distances, which is what we want to minimize, is

$$f(m, b) = (2 - (b - m))^2 + (-1 - b)^2 + (1 - (b + m))^2.$$

To find the critical points, we compute the partial derivatives with respect to m and b,

$$\begin{aligned}
f_m &= 2(2 - b + m) + 0 + 2(1 - b - m)(-1) \\
&= 4 - 2b + 2m - 2 + 2b + 2m \\
&= 2 + 4m, \\
f_b &= 2(2 - b + m)(-1) + 2(-1 - b)(-1) + 2(1 - b - m)(-1) \\
&= -4 + 2b - 2m + 2 + 2b - 2 + 2b + 2m \\
&= -4 + 6b.
\end{aligned}$$

Setting both partial derivatives equal to zero, we get a system of equations:

$$2 + 4m = 0,$$
$$-4 + 6b = 0.$$

The solution is $m = -1/2$ and $b = 2/3$. You can check that it is a minimum. Hence, the regression line is $y = \dfrac{2}{3} - \dfrac{1}{2}x$.

5. We have $\sum x_i = 6$, $\sum y_i = 5$, $\sum x_i^2 = 14$, and $\sum y_i x_i = 12$. Thus

$$b = (14 \cdot 5 - 6 \cdot 12) / \left(3 \cdot 14 - 6^2\right) = -1/3.$$
$$m = (3 \cdot 12 - 6 \cdot 5) / \left(3 \cdot 14 - 6^2\right) = 1.$$

The line is $y = x - \dfrac{1}{3}$, which agrees with the answer to Example 1.

CHAPTER NINE

Solutions for Section 9.1

1. (a) = (III), (b) = (IV), (c) = (I), (d) = (II).

5. The rate of change of Q is proportional to Q so we have

$$\frac{dQ}{dt} = kQ,$$

for some constant k. Since the radioactive substance is decaying, the quantity present, Q, is decreasing. The derivative dQ/dt must be negative, so the constant of proportionality k is negative.

9. The amount of alcohol, A, is decreasing at a constant rate of 1 ounce per hour, so we have

$$\frac{dA}{dt} = -1.$$

The negative sign indicates that the amount of alcohol is decreasing.

13. (a) To see if W is increasing or decreasing, we determine whether the derivative dW/dt is positive or negative. When $W = 10$, we have

$$\frac{dW}{dt} = 5W - 20 = 5(10) - 20 = 30 > 0.$$

Since dW/dt is positive when $W = 10$, the quantity W is increasing.
When $W = 2$, we have

$$\frac{dW}{dt} = 5W - 20 = 5(2) - 20 = -10 < 0.$$

Since dW/dt is negative when $W = 2$, the quantity W is decreasing.

(b) We set $dW/dt = 0$ and solve:

$$\frac{dW}{dt} = 0$$
$$5W - 20 = 0$$
$$W = 4.$$

The rate of change of W is zero when $W = 4$.

Solutions for Section 9.2

1. (a) Since $y = x^2$, we have $y' = 2x$. Substituting these functions into our differential equation, we have

$$xy' - 2y = x(2x) - 2(x^2) = 2x^2 - 2x^2 = 0.$$

Therefore, $y = x^2$ is a solution to the differential equation $xy' - 2y = 0$.

(b) For $y = x^3$, we have $y' = 3x^2$. Substituting gives:

$$xy' - 2y = x(3x^2) - 2(x^3) = 3x^3 - 2x^3 = x^3.$$

Since x^3 does not equal 0 for all x, we see that $y = x^3$ is not a solution to the differential equation.

5. Since $dy/dx = 0.1$, the slope of the curve is 0.1 at all points. Thus the curve is a line with positive slope, such as Graph F.

9. Since $dy/dx = y$, the slope of the solution curve will be positive for positive y-values and negative for negative y-values. In addition, the slope will be bigger for large y and closer to zero when the y-value is closer to zero. A possible solution curve for this differential equation is Graph A.

13. We know that at time $t = 0$, the value of y is 8. Since we are told that $dy/dt = 4 - y$, we know that at time $t = 0$

$$\frac{dy}{dt} = 4 - 8 = -4.$$

As t goes from 0 to 1, y will decrease by 4, so at $t = 1$,

$$y = 8 - 4 = 4$$

Likewise, we get that at $t = 1$,

$$\frac{dy}{dt} = 4 - 4 = 0$$

so that at $t = 2$,

$$y = 4 + 0(1) = 4.$$

At $t = 2$, $\dfrac{dy}{dt} = 4 - 4 = 0$ so that at $t = 3$, $y = 4 + 0 = 4$.

At $t = 3$, $\dfrac{dy}{dt} = 4 - 4 = 0$ so that at $t = 4$, $y = 4 + 0 = 4$.

Thus we get the following table

t	0	1	2	3	4
y	8	4	4	4	4

17. At $t = 0$, we know $P = 70$ and we can compute the value of dP/dt:

$$\text{At } t = 0, \quad \text{we have} \quad \frac{dP}{dt} = 0.2P - 10 = 0.2(70) - 10 = 4.$$

The population is increasing at a rate of 4 million fish per year. At the end of the first year, the fish population will have grown by about 4 million fish, and so we have:

$$\text{At } t = 1, \quad \text{we estimate} \quad P = 70 + 4 = 74.$$

We can now use this new value of P to calculate dP/dt at $t = 1$:

$$\text{At } t = 1, \quad \text{we have} \quad \frac{dP}{dt} = 0.2P - 10 = 0.2(74) - 10 = 4.8,$$

and so:

$$\text{At } t = 2, \quad \text{we estimate} \quad P = 74 + 4.8 = 78.8.$$

Continuing in this way, we obtain the values in Table 9.1.

Table 9.1

t	0	1	2	3
P	70	74	78.8	84.56

21. Since $y = x^2 + k$ we know that

$$y' = 2x.$$

Substituting $y = x^2 + k$ and $y' = 2x$ into the differential equation we get

$$\begin{aligned}
10 &= 2y - xy' \\
&= 2(x^2 + k) - x(2x) \\
&= 2x^2 + 2k - 2x^2 \\
&= 2k
\end{aligned}$$

Thus, $k = 5$ is the only solution.

Solutions for Section 9.3

1. See Figure 9.1. Other choices of solution curves are, of course, possible.

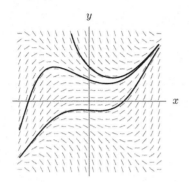
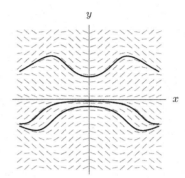

Figure 9.1

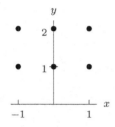

Figure 9.2

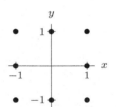

Figure 9.3

5. (a) See Figure 9.4.

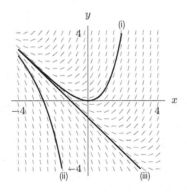

Figure 9.4

(b) The solution through $(-1, 0)$ appears to be linear with equation $y = -x - 1$.

(c) If $y = -x - 1$, then $y' = -1$ and $x + y = x + (-x - 1) = -1$, so this checks as a solution.

9. III. The slope field appears to be near zero at $P = 1$ and $P = 0$, so this rules out $dP/dt = P - 1$, which has a slope of -1 at $P = 0$. Between $P = 0$ and $P = 1$, the slopes in the figure are positive, so this rules out $dP/dt = P(P - 1)$, which has negative values for $0 < P < 1$. To decide between the remaining two possibilities, note that when $P = 1/2$, the slopes in the figure appear to be about 1. The differential equation $dP/dt = 3P(1 - P)$ gives a slope of $3 \cdot 1/2 \cdot (1 - 1/2) = 3/4$, while the differential equation $dP/dt = 1/3P(1 - P)$ gives a slope of $1/12$ which is clearly too small. Thus the best answer is $dP/dt = 3P(1 - P)$.

13. As $x \to \infty$, $y \to \infty$, no matter what the starting point is.

17. When $a = 1$ and $b = 2$, the Gompertz equation is $y' = -y \ln(y/2) = y \ln(2/y) = y(\ln 2 - \ln y)$. This differential equation is similar to the differential equation $y' = y(2 - y)$ in certain ways. For example, in both equations y' is positive for $0 < y < 2$ and negative for $y > 2$. Also, for y-values close to 2, the quantities $(\ln 2 - \ln y)$ and $(2 - y)$ are both close to 0, so $y(\ln 2 - \ln y)$ and $y(2 - y)$ are approximately equal to zero. Thus around $y = 2$ the slope fields look almost the same. This happens again around $y = 0$, since around $y = 0$ both $y(2 - y)$ and $y(\ln 2 - \ln y)$ go to 0. Finally, for $y > 2$, $\ln y$ grows much slower than y, so the slope field for $y' = y(\ln 2 - \ln y)$ is less steep, negatively, than for $y' = y(2 - y)$.

Solutions for Section 9.4

1. The equation given is in the form

$$\frac{dP}{dt} = kP.$$

Thus we know that the general solution to this equation will be

$$P = Ce^{kt}.$$

And in our case, with $k = 0.02$ and $C = 20$ we get

$$P = 20e^{0.02t}.$$

5. The equation is in the form $dp/dq = kp$, so the general solution is the exponential function

$$p = Ce^{-0.1q}.$$

We find C using the condition that $p = 100$ when $q = 5$.

$$p = Ce^{-0.1q}$$
$$100 = Ce^{-0.1(5)}$$
$$C = \frac{100}{e^{-0.5}} = 164.87.$$

The solution is

$$p = 164.87e^{-0.1q}.$$

9. **(a)** If $B = f(t)$ (where t is in years)

$$\frac{dB}{dt} = \text{Rate at which interest is earned} + \text{Rate at which money is deposited}$$
$$= 0.10B + 1000.$$

(b)

$$\frac{dB}{dt} = 0.1(B + 10{,}000)$$

We know that a differential equation of the form

$$\frac{dB}{dt} = k(B - A)$$

has general solution:

$$B = Ce^{kt} + A.$$

Thus, in our case

$$B = Ce^{0.1t} - 10{,}000.$$

For $t = 0$, $B = 0$, hence $C = 10{,}000$. Therefore, $B = 10{,}000e^{0.1t} - 10{,}000$.

13. (a) Since the amount leaving the blood is proportional to the quantity in the blood,

$$\frac{dQ}{dt} = -kQ \quad \text{for some positive constant } k.$$

Thus $Q = Q_0 e^{-kt}$, where Q_0 is the initial quantity in the bloodstream. Only 20% is left in the blood after 3 hours. Thus $0.20 = e^{-3k}$, so $k = \frac{\ln 0.20}{-3} \approx 0.5365$. Therefore $Q = Q_0 e^{-0.5365t}$.

(b) Since 20% is left after 3 hours, after 6 hours only 20% of that 20% will be left. Thus after 6 hours only 4% will be left, so if the patient is given 100 mg, only 4 mg will be left 6 hours later.

17. (a) Assuming that the world's population grows exponentially, satisfying $dP/dt = cP$, and that the land in use for crops is proportional to the population, we expect A to satisfy $dA/dt = kA$.

(b) We have $A(t) = A_0 e^{kt} = 4.55 \cdot 10^9 e^{kt}$, where t is the number of years after 1966. Since 30 years later the amount of land in use is 4.93 billion hectares, we have

$$4.93 \cdot 10^9 = (4.55 \cdot 10^9) e^{k(30)},$$

so

$$e^{30k} = \frac{4.93}{4.55}.$$

Solving for k gives

$$k = \frac{\ln (4.93/4.55)}{30} = 0.00267.$$

Thus,

$$A = (4.55 \cdot 10^9) e^{0.00267t}.$$

We want to find t such that

$$6 \cdot 10^9 = A(t) = (4.55 \cdot 10^9) e^{0.00267t}.$$

Taking logarithms gives

$$t = \frac{\ln (6/4.55)}{0.00267} = 103.608 \text{ years.}$$

This model predicts land will have run out 104 years after 1966, that is by the year 2070.

Solutions for Section 9.5

1. We know that the general solution to a differential equation of the form

$$\frac{dy}{dt} = k(y - A)$$

is

$$y = A + Ce^{kt}.$$

Thus in our case we get

$$y = 200 + Ce^{0.5t}.$$

We know that at $t = 0$ we have $y = 50$, so solving for C we get

$$y = 200 + Ce^{0.5t}$$
$$50 = 200 + Ce^{0.5(0)}$$
$$-150 = Ce^0$$
$$C = -150.$$

Thus we get

$$y = 200 - 150e^{0.5t}.$$

5. We know that the general solution to a differential equation of the form

$$\frac{dB}{dt} = k(B - A)$$

is

$$B = A + Ce^{kt}.$$

To get our equation in this form we factor out a 4 to get

$$\frac{dB}{dt} = 4\left(B - \frac{100}{4}\right) = 4(B - 25).$$

Thus in our case we get

$$B = Ce^{4t} + 25.$$

We know that at $t = 0$ we have $B = 20$, so solving for C we get

$$B = 25 + Ce^{4t}$$
$$20 = 25 + Ce^{4(0)}$$
$$-5 = Ce^0$$
$$C = -5.$$

Thus we get

$$B = 25 - 5e^{4t}.$$

9. In order to check that $y = A + Ce^{kt}$ is a solution to the differential equation

$$\frac{dy}{dt} = k(y - A),$$

we must show that the derivative of y with respect to t is equal to $k(y - A)$:

$$y = A + Ce^{kt}$$
$$\frac{dy}{dt} = 0 + (Ce^{kt})(k)$$
$$= kCe^{kt}$$
$$= k(Ce^{kt} + A - A)$$
$$= k\left((Ce^{kt} + A) - A\right)$$
$$= k(y - A)$$

13. (a) We know that the rate by which the account changes every year is

$$\text{Rate of change of balance} = \text{Rate of increase} - \text{Rate of decrease}.$$

Since $1000 will be withdrawn every year, we know that the account decreases by $1000 every year. We also know that the account accumulates interest at 7% compounded continuously. Thus the amount by which the account increases each year is

$$\text{Rate balance increases per year} = 7\%(\text{Account balance}) = 0.07(\text{Account balance}).$$

Denoting the account balance by B we get

$$\text{Rate balance increases per year} = 0.07B.$$

Thus we get

$$\text{Rate of change of balance} = 0.07B - 1000.$$

or

$$\frac{dB}{dt} = 0.07B - 1000,$$

with t measured in years.

(b) The equilibrium solution makes the derivative 0, so

$$\frac{dB}{dt} = 0$$
$$0.07B - 1000 = 0$$
$$B = \frac{1000}{0.07} \approx \$14{,}285.71.$$

(c) We know that the general solution to a differential equation of the form

$$\frac{dB}{dt} = k(B - A)$$

is

$$B = Ce^{kt} + A.$$

To get our equation in this form we factor out a 0.07 to get

$$\frac{dB}{dt} = 0.07\left(B - \frac{1000}{0.07}\right) \approx 0.07(B - 14{,}285.71).$$

Thus in our case we get

$$B = Ce^{0.07t} + 14{,}285.71.$$

We know that at $t = 0$ we have $B = \$10{,}000$ so solving for C we get

$$B = Ce^{0.07t} + 14{,}285.71$$
$$10{,}000 = Ce^{4(0)} + 14{,}285.71$$
$$-4285.71 = Ce^{0}$$
$$C = -4285.71.$$

Thus we get

$$B = 14{,}285.71 - (4285.71)e^{0.07t}.$$

(d) Substituting the value $t = 5$ into our function for B we get

$$B(t) = 14{,}285.71 - (4285.71)e^{0.07t}$$
$$B(5) = 14{,}285.71 - (4285.71)e^{0.07(5)}$$
$$= 14{,}285.71 - (4285.71)e^{0.35}$$
$$B(5) \approx \$8204$$

(e) From Figure 9.5 we see that in the long run there is no money left in the account.

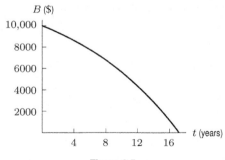

Figure 9.5

17. (a) We have $\dfrac{dy}{dt} = -k(y - a)$, where $k > 0$ and a are constants.

(b) We know that the general solution to a differential equation of the form

$$\frac{dy}{dt} = -k(y - a)$$

is

$$y = Ce^{-kt} + a.$$

We can assume that right after the course is over (at $t = 0$) 100% of the material is remembered. Thus we get

$$y = Ce^{-kt} + a$$
$$1 = Ce^0 + a$$
$$C = 1 - a.$$

Thus

$$y = (1 - a)e^{-kt} + a.$$

(c) As $t \to \infty$, $e^{-kt} \to 0$, so $y \to a$.

Thus, a represents the fraction of material which is remembered in the long run. The constant k tells us about the rate at which material is forgotten.

21. (a) We know that the general solution to a differential equation of the form

$$\frac{dH}{dt} = -k(H - 200)$$

is

$$H = Ce^{-kt} + 200.$$

We know that at $t = 0$ we have $H = 20$ so solving for C we get

$$H = Ce^{-kt} + 200$$
$$20 = Ce^0 + 200$$
$$C = -180.$$

Thus we get

$$H = -180e^{-kt} + 200.$$

(b) Using part (a), we have $120 = 200 - 180e^{-k(30)}$. Solving for k, we have $e^{-30k} = \frac{-80}{-180}$, giving

$$k = \frac{\ln \frac{4}{9}}{-30} \approx 0.027.$$

Note that this k is correct if t is given in *minutes*. (If t is given in hours, $k = \frac{\ln \frac{4}{9}}{-\frac{1}{2}} \approx 1.62$.)

25. Differentiate with respect to t on both sides of the equation:

$$\frac{y - A}{y_0 - A} = e^{kt}.$$

Since A and y_0 are constant, we have

$$\frac{y'}{y_0 - A} = ke^{kt} = k\frac{y - A}{y_0 - A}.$$

Multiplication by $y_0 - A$ gives

$$y' = k(y - A)$$

which shows that y satisfies the given differential equation.

Now we need to show that the initial condition, $y(0) = y_0$, is satisfied. Substituting $t = 0$ gives

$$\frac{y(0) - A}{y_0 - A} = e^{k \cdot 0} = 1$$
$$y(0) - A = y_0 - A$$
$$y(0) = y_0$$

which shows that y satisfies the given initial condition.

Solutions for Section 9.6

1. (a) The x population is unaffected by the y population—it grows exponentially no matter what the y population is, even if $y = 0$. If alone, the y population decreases to zero exponentially, because its equation becomes $dy/dt = -0.1y$.
    (b) Here, interaction between the two populations helps the y population but does not effect the x population. This is not a predator-prey relationship; instead, this is a one-way relationship, where the y population is helped by the existence of x's. These equations could, for instance, model the interaction of rhinoceroses (x) and dung beetles (y).

5. $\dfrac{dx}{dt} = -x - xy, \quad \dfrac{dy}{dt} = -y - xy$

9. This is an example of a predator-prey relationship. Normally, we would expect the worm population, in the absence of predators, to increase without bound. As the number of worms w increases, so would the rate of increase dw/dt; in other words, the relation $dw/dt = w$ might be a reasonable model for the worm population in the absence of predators.

    However, since there are predators (robins), dw/dt won't be that big. We must lessen dw/dt. It makes sense that the more interaction there is between robins and worms, the more slowly the worms are able to increase their numbers. Hence we lessen dw/dt by the amount wr to get $dw/dt = w - wr$. The term $-wr$ reflects the fact that more interactions between the species means slower reproduction for the worms.

    Similarly, we would expect the robin population to decrease in the absence of worms. We'd expect the population decrease at a rate related to the current population, making $dr/dt = -r$ a reasonable model for the robin population in absence of worms. The negative term reflects the fact that the greater the population of robins, the more quickly they are dying off. The wr term in $dr/dt = -r + wr$ reflects the fact that the more interactions between robins and worms, the greater the tendency for the robins to increase in population.

13. Sketching the trajectory through the point $(2, 2)$ on the slope field given shows that the maximum robin population is about 2500, and the minimum robin population is about 500. When the robin population is at its maximum, the worm population is about 1,000,000.

17. (a) Substituting $w = 2.2$ and $r = 1$ into the differential equations gives

$$\frac{dw}{dt} = 2.2 - (2.2)(1) = 0$$

$$\frac{dr}{dt} = -1 + 1(2.2) = 1.2.$$

    (b) Since the rate of change of w with time is 0,

$$\text{At } t = 0.1, \text{ we estimate } w = 2.2$$

    Since the rate of change of r is 1.2 thousand robins per unit time,

$$\text{At } t = 0.1, \text{ we estimate } r = 1 + 1.2(0.1) = 1.12 \approx 1.1.$$

    (c) We must recompute the derivatives. At $t = 0.1$, we have

$$\frac{dw}{dt} = 2.2 - 2.2(1.12) = -0.264$$

$$\frac{dr}{dt} = -1.12 + 1.12(2.2) = 1.344.$$

    Then at $t = 0.2$, we estimate

$$w = 2.2 - 0.264(0.1) = 2.1736 \approx 2.2$$

$$r = 1.12 + 1.344(0.1) = 1.2544 \approx 1.3$$

    Recomputing the derivatives at $t = 0.2$ gives

$$\frac{dw}{dt} = 2.1736 - 2.1736(1.2544) = -0.553$$

$$\frac{dr}{dt} = -1.2544 + 1.2544(2.1736) = 1.472$$

    Then at $t = 0.3$, we estimate

$$w = 2.1736 - 0.553(0.1) = 2.1183 \approx 2.1$$

$$r = 1.2544 + 1.472(0.1) = 1.4016 \approx 1.4.$$

21. Here x and y both increase at about the same rate.

25. (a) The rate dQ_1/dt is the sum of three terms that represent the three changes in Q_1:

$$\frac{dQ_1}{dt} = A - k_1 Q_1 + k_2 Q_2.$$

The term A is the rate at which Q_1 increases due to creation of new toxin.

The term $-k_1 Q_1$ is the rate at which Q_1 decreases due to flow of toxin into the blood. The constant k_1 is a positive constant of proportionality.

The term $k_2 Q_2$ is the rate at which Q_1 increases due to flow of toxin out of the blood. The constant k_2 is a second positive constant.

(b) The rate dQ_2/dt is the sum of three terms that represent the three changes in Q_2:

$$\frac{dQ_2}{dt} = -k_3 Q_2 + k_1 Q_1 - k_2 Q_2.$$

The term $-k_3 Q_2$ is the rate at which Q_2 decreases due to removal of toxin by dialysis. The constant k_3 is a positive constant of proportionality.

The term $k_1 Q_1$ is the rate at which Q_2 increases due to flow of toxin into the blood. The constant k_1 is the same positive constant as in part (a).

The term $-k_2 Q_2$ is the rate at which Q_2 decreases due to flow of toxin out of the blood. The constant k_2 is the same positive constant as in part (a).

Solutions for Section 9.7

1. Susceptible people are infected at a rate proportional to the product of S and I. As susceptible people become infected, S decreases at a rate of aSI and (since these same people are now infected) I increases at the same rate. At the same time, infected people are recovering at a rate proportional to the number infected, so I is decreasing at a rate of bI.

5. (a) $I_0 = 1,\ S_0 = 349$

(b) Since $\dfrac{dI}{dt} = 0.0026SI - 0.5I = 0.0026(349)(1) - 0.5(1) > 0$, so I is increasing. The number of infected people will increase, and the disease will spread. This is an epidemic.

9. The threshold value of S is the value at which I is a maximum. When I is a maximum,

$$\frac{dI}{dt} = 0.04SI - 0.2I = 0,$$

so

$$S = 0.2/0.04 = 5.$$

Solutions for Chapter 9 Review

1. (a) (III) An island can only sustain the population up to a certain size. The population will grow until it reaches this limiting value.

(b) (V) The ingot will get hot and then cool off, so the temperature will increase and then decrease.

(c) (I) The speed of the car is constant, and then decreases linearly when the breaks are applied uniformly.

(d) (II) Carbon-14 decays exponentially.

(e) (IV) Tree pollen is seasonal, and therefore cyclical.

5. (a) Slope field I corresponds to $\dfrac{dy}{dx} = 1 + y$ and slope field II corresponds to $\dfrac{dy}{dx} = 1 + x$.

(b) See Figures 9.6 and 9.7.

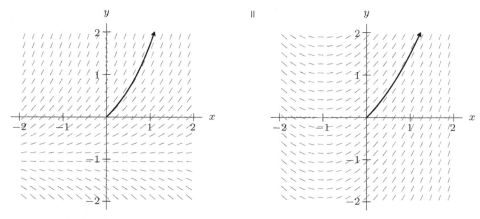

Figure 9.6: $\frac{dy}{dx} = 1 + y$ **Figure 9.7:** $\frac{dy}{dx} = 1 + x$

(c) Slope field I has an equilibrium solution at $y = -1$, since $\frac{dy}{dx} = 0$ at $y = -1$. We see in the slope field that this equilibrium solution is unstable. Slope field II does not have any equilibrium solutions.

9. (a) We know that the balance, B, increases at a rate proportional to the current balance. Since interest is being earned at a rate of 7% compounded continuously we have

$$\text{Rate at which interest is earned} = 7\% \text{ (Current balance)}$$

or in other words, if t is time in years,

$$\frac{dB}{dt} = 7\%(B) = 0.07B.$$

(b) The equation is in the form

$$\frac{dB}{dt} = kB$$

so we know that the general solution will be

$$B = B_0 e^{kt}$$

where B_0 is the value of B when $t = 0$, i.e., the initial balance. In our case we have $k = 0.07$ so we get

$$B = B_0 e^{0.07t}.$$

(c) We are told that the initial balance, B_0, is \$5000 so we get

$$B = 5000 e^{0.07t}.$$

(d) Substituting the value $t = 10$ into our formula for B we get

$$B = 5000 e^{0.07t}$$
$$B(10) = 5000 e^{0.07(10)}$$
$$= 5000 e^{0.7}$$
$$B(10) \approx \$10{,}068.76$$

13. Integrating both sides we get

$$y = \frac{5}{2} t^2 + C,$$

where C is a constant.

17. We know that the general solution to the differential equation

$$\frac{dP}{dt} = k(P - A)$$

is

$$P = Ce^{kt} + A.$$

Thus in our case we factor out -2 to get

$$\frac{dP}{dt} = -2\left(P + \frac{10}{-2}\right) = -2(P - 5).$$

Thus the general solution to our differential equation is

$$P = Ce^{-2t} + 5,$$

where C is some constant.

21. We find the temperature of the orange juice as a function of time. Newton's Law of Heating says that the rate of change of the temperature is proportional to the temperature difference. If S is the temperature of the juice, this gives us the equation

$$\frac{dS}{dt} = -k(S - 65) \qquad \text{for some constant } k.$$

Notice that the temperature of the juice is increasing, so the quantity dS/dt is positive. In addition, $S = 40$ initially, making the quantity $(S - 65)$ negative.
We know that the general solution to a differential equation of the form

$$\frac{dS}{dt} = -k(S - 65)$$

is

$$S = Ce^{-kt} + 65.$$

Since at $t = 0$, $S = 40$, we have $40 = 65 + C$, so $C = -25$. Thus, $S = 65 - 25e^{-kt}$ for some positive constant k. See Figure 9.8 for the graph.

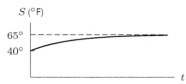

Figure 9.8: Graph of
$S = 65 - 25e^{-kt}$ for $k > 0$

25. (a)

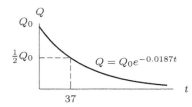

(b) $\dfrac{dQ}{dt} = -kQ$

(c) Since $25\% = 1/4$, it takes two half-lives $= 74$ hours for the drug level to be reduced to 25%. Alternatively, $Q = Q_0e^{-kt}$ and $\frac{1}{2} = e^{-k(37)}$, we have

$$k = -\frac{\ln(1/2)}{37} \approx 0.0187.$$

Therefore $Q = Q_0e^{-0.0187t}$. We know that when the drug level is 25% of the original level that $Q = 0.25Q_0$. Setting these equal, we get

$$0.25 = e^{-0.0187t}.$$

giving

$$t = -\frac{\ln(0.25)}{0.0187} \approx 74 \text{ hours} \approx 3 \text{ days}.$$

29. **(a)** Since the rate of change of the weight is given by

$$\frac{dW}{dt} = \frac{1}{3500}(\text{Intake} \ - \ \text{Amount to maintain weight})$$

we have

$$\frac{dW}{dt} = \frac{1}{3500}(I - 20W).$$

(b) To find the equilibrium, we solve $dW/dt = 0$, or

$$\frac{1}{3500}(I - 20W) = 0.$$

Solving for W, we get $W = I/20$.

   This means that if an athletic adult male weighing $I/20$ pounds has a constant caloric intake of I Calories per day, his weight remains constant. We expect the equilibrium to be stable because an athletic adult male slightly over the equilibrium weight loses weight because his caloric intake is too low to maintain the higher weight. Similarly, an adult male slightly under the equilibrium weight gains weight because his caloric intake is higher than required to maintain his weight.

(c) We know that the general solution to a differential equation of the form

$$\frac{dW}{dt} = k(W - A)$$

is

$$W = Ce^{kt} + A.$$

Factoring out a -20 on the left side we get

$$\frac{dW}{dt} = \frac{-20}{3500}\left(W - \frac{-I}{-20}\right) = -\frac{2}{350}\left(W - \frac{I}{20}\right).$$

Thus in our case we get

$$W = Ce^{-\frac{2}{350}t} + \frac{I}{20}.$$

Let us call the person's initial weight W_0 at $t = 0$. Then $W_0 = I/20 + Ce^0$, so $C = W_0 - I/20$. Thus

$$W = \frac{I}{20} + \left(W_0 - \frac{I}{20}\right)e^{-(1/175)t}.$$

(d) Using part (c), we have $W = 150 + 10e^{-(1/175)t}$. This means that $W \to 150$ as $t \to \infty$. See Figure 9.9.

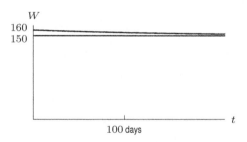

Figure 9.9

STRENGTHEN YOUR UNDERSTANDING

1. True, since dQ/dt represents the rate of change of Q.

5. True, since when two quantities are proportional, one is a constant times the other.

9. True. The rate the drug is entering the body is 12 mg per hour and the rate the drug is leaving the body is $0.063Q$ mg per hour.

13. False, since substituting $P = 5$ on the left-hand side of the differential equation gives 0, but on the right-hand side gives 75.

17. False. Since $y' = 0.2y$, when $y = 100$ we have $y' = 0.2 \cdot 100 = 20$. The variable y is changing at a rate of 20 units per unit of time. This tells us that y increases approximately 20 units between $t = 0$ and $t = 1$, so we expect $y(1) \approx 100 + 20 = 120$.

21. False, since when $x = 3$, we have $dy/dx = 2 \cdot 3 = 6$.

25. True, since when $x = 3$ and $y = 2$, we have $dy/dx = 3 \cdot 3 \cdot 2 = 18$.

29. False, since when $P > 3$ we have $12 - 4P < 0$ so $dP/dt < 0$.

33. False. That solution would be the solution to the differential equation $dw/dr = 0.3w$.

37. False, the correct differential equation is $dB/dt = 0.03B$.

41. True, as explained in the text.

45. False. The initial condition gives $C = 10$ so the correct solution is $A = 40 + 10e^{0.25t}$.

49. True, since setting $dH/dt = 0$ gives $H = 225$.

53. True. Since X has a negative impact on Y, the coefficient of the interaction term must be negative.

57. True, since the coefficients -0.15 and -0.18 of the interaction terms are negative.

61. False. It is negative because people in the susceptible group are becoming sick and moving to the group of infected people.

65. False. The parameter a will be larger for Type I flu.

69. True. To see if I will increase, we see if dI/dt is positive. We have $dI/dt = 0.001SI - 0.3I = 0.001(500)(100) - 0.3(100) = 50 - 30 = 20 > 0$.

Solutions to Problems on Separation of Variables

1. Separating variables gives

$$\int \frac{1}{P}dP = -\int 2dt,$$

so

$$\ln |P| = -2t + C.$$

Therefore

$$P = \pm e^{-2t+C} = Ae^{-2t}.$$

The initial value $P(0) = 1$ gives $1 = A$, so

$$P = e^{-2t}.$$

5. Separating variables gives

$$\int \frac{1}{u^2} du = \int \frac{1}{2} dt$$

or

$$-\frac{1}{u} = \frac{1}{2}t + C.$$

The initial condition gives $C = -1$ and so

$$u = \frac{1}{1 - (1/2)t}.$$

9. Separating variables gives

$$\frac{dz}{dt} = te^z$$

$$e^{-z}dz = tdt$$

$$\int e^{-z}\,dz = \int t\,dt,$$

so

$$-e^{-z} = \frac{t^2}{2} + C.$$

Since the solution passes through the origin, $z = 0$ when $t = 0$, we must have

$$-e^{-0} = \frac{0}{2} + C, \text{ so } C = -1.$$

Thus

$$-e^{-z} = \frac{t^2}{2} - 1,$$

or

$$z = -\ln\left(1 - \frac{t^2}{2}\right).$$

13. (a) Yes   **(b)** No   **(c)** Yes
   (d) No   **(e)** Yes   **(f)** Yes
   (g) No   **(h)** Yes   **(i)** No
   (j) Yes   **(k)** Yes   **(l)** No

17. Factoring and separating variables gives

$$\frac{dR}{dt} = a\left(R + \frac{b}{a}\right)$$

$$\int \frac{dR}{R + b/a} = \int a\, dt$$

$$\ln\left|R + \frac{b}{a}\right| = at + C$$

$$R = -\frac{b}{a} + Ae^{at}, \quad \text{where } A \text{ can be any constant.}$$

21. (a) The slope field for $dy/dx = xy$ is in Figure 9.10.

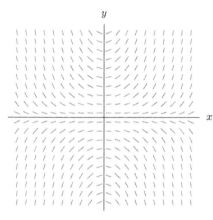

Figure 9.10

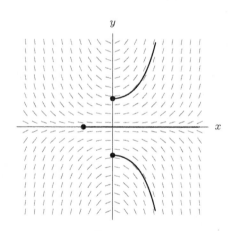

Figure 9.11

(b) Some solution curves are shown in Figure 9.11.
(c) Separating variables gives

$$\int \frac{1}{y}\, dy = \int x\, dx$$

or

$$\ln|y| = \frac{1}{2}x^2 + C.$$

Solving for y gives

$$y(x) = Ae^{x^2/2}$$

where $A = \pm e^C$. In addition, $y(x) = 0$ is a solution. So $y(x) = Ae^{x^2/2}$ is a solution for any A.

CHAPTER TEN

Solutions for Section 10.1

1. Adding the terms, we see that

$$3 + 3 \cdot 2 + 3 \cdot 2^2 = 3 + 6 + 12 = 21.$$

We can also find the sum using the formula for a finite geometric series with $a = 3$, $r = 2$, and $n = 3$:

$$3 + 3 \cdot 2 + 3 \cdot 2^2 = \frac{3(1 - 2^3)}{1 - 2} = 3(8 - 1) = 21.$$

5. Yes, $a = 2$, ratio $= 1/2$.

9. Yes, $a = y^2$, ratio $= y$.

13. This is an infinite geometric series with $a = 75$ and $r = 0.22$. Since $-1 < r < 1$, the series converges and the sum is given by:

$$\text{Sum} = 75 + 75(0.22) + 75(0.22)^2 + \cdots = \frac{75}{1 - 0.22} = 96.154.$$

17. This is a finite geometric series with $a = 3$, $r = 1/2$, and $n - 1 = 10$, so $n = 11$. Thus

$$\text{Sum} = \frac{3(1 - (1/2)^{11})}{1 - 1/2} = 3 \cdot 2 \frac{(2^{11} - 1)}{2^{11}} = \frac{3(2^{11} - 1)}{2^{10}}.$$

21. This is an infinite geometric series with $a = -2$ and $r = -1/2$. Since $-1 < r < 1$, the series converges and

$$\text{Sum} = \frac{-2}{1 - (-1/2)} = -\frac{4}{3}.$$

25. (a) Notice that the 6th deposit is made 5 months after the first deposit, so the first deposit has grown to $500(1.001)^5$ at that time. The balance in the account right after the 6th deposit is the sum

$$\text{Balance} = 500 + 500(1.001) + 500(1.001)^2 + \cdots + 500(1.001)^5.$$

We find the sum using the formula for a finite geometric series with $a = 500$, $r = 1.001$, and $n = 6$:

$$\text{Balance right after } 6^{\text{th}} \text{ deposit} = \frac{500(1 - (1.001)^6)}{1 - 1.001} = \$3007.51.$$

Since each deposit is $500, the balance in the account right before the 6th deposit is $3007.51 - 500 = \$2507.51$.

(b) Similarly, the 12th deposit is made 11 months after the first deposit, so the first deposit has grown to $500(1.001)^{11}$ at that time. The balance in the account right after the 12th deposit is the sum

$$\text{Balance} = 500 + 500(1.001) + 500(1.001)^2 + \cdots + 500(1.001)^{11}.$$

We find the sum using the formula for a finite geometric series with $a = 500$, $r = 1.001$, and $n = 12$:

$$\text{Balance right after } 12^{\text{th}} \text{ deposit} = \frac{500(1 - (1.001)^{12})}{1 - 1.001} = \$6033.11.$$

Since each deposit is $500, the balance in the account right before the 12th deposit is $6033.11 - 500 = \$5533.11$.

29. (a) The average quantity in the body is $(65 + 15)/2 = 40$ mg.

(b) The average concentration for this patient (in milligrams of quinine per kilogram of body weight) is $(40 \text{ mg})/(70 \text{ kg})$ = 0.57 mg/kg. This average concentration falls within the range that is both safe and effective.

(c) (i) Since this treatment produces an average of 40 mg of quinine in the body, a body weight W kg produces an average concentration below 0.4 mg/kg if

$$\frac{40}{W} < 0.4$$

so

$$W > 100.$$

The treatment is not effective for anyone weighing more than 100 kg (or about 220 pounds.)

(ii) A body weight W kg produces an average concentration above 3.0 mg/kg if

$$\frac{40}{W} > 3.0$$

so

$$W < 13.3.$$

The treatment is unsafe for anyone weighing less than 13.3 kg (or about 30 pounds.)

Solutions for Section 10.2

1. The 20th deposit has just been made and has not yet earned any interest, the 19th deposit has earned interest for one year and is worth $1000(1.02)$, the 18th deposit has earned interest for two years and is worth $1000(1.02)^2$, and so on. The first deposit has earned interest for 19 years and is worth $1000(1.02)^{19}$. The total balance in the account right after the 20th deposit is

$$\text{Balance } = 1000 + 1000(1.02) + 1000(1.02)^2 + \cdots + 1000(1.02)^{19}.$$

This is a finite geometric series with $a = 1000$, $r = 1.02$, and $n = 20$. Using the formula for the sum of a finite geometric series, we have

$$\text{Balance after the } 20^{\text{th}} \text{ deposit } = \frac{1000(1 - (1.02)^{20})}{1 - 1.02} = \$24{,}297.37.$$

The twenty annual deposits of $1000 have contributed a total of $20,000 to this balance, and the remaining $4297.37 in the account comes from the interest earned.

5. We sum the geometric series

$$\text{Present value } = 20{,}000 + \frac{20{,}000}{1.01} + \frac{20{,}000}{1.01^2} + \cdots$$
$$= \frac{20{,}000}{1 - 1/1.01} = 2{,}020{,}000.$$

Thus, the present value is 2.02 million dollars.

9. You earn 1 cent the first day, 2 cents the second day, $2^2 = 4$ the third day, $2^3 = 8$ the fourth day, and so on. On the nth day, you earn 2^{n-1} cents. We have

$$\text{Total earnings for } n \text{ days } = 1 + 2 + 2^2 + 2^3 + \cdots + 2^{n-1}.$$

This is a finite geometric series with $a = 1$ and $r = 2$. We use the formula for the sum:

$$\text{Total earnings for } n \text{ days } = \frac{1 - 2^n}{1 - 2} = 2^n - 1.$$

(a) Using $n = 7$, we see that

$$\text{Total earnings for 7 days } = 2^7 - 1 = 127 = \$1.27.$$

(b) Using $n = 14$, we see that

$$\text{Total earnings for 14 days } = 2^{14} - 1 = 16383 = \$163.83.$$

(c) Using $n = 21$, we see that

$$\text{Total earnings for 21 days } = 2^{21} - 1 = 2097151 = \$20{,}971.51.$$

(d) Using $n = 28$, we see that

$$\text{Total earnings for 28 days } = 2^{28} - 1 = 268435455 = \$2{,}684{,}354.55.$$

13.

$$\text{Present value of first coupon} = \frac{50}{1.04}$$

$$\text{Present value of second coupon} = \frac{50}{(1.04)^2}, \text{etc.}$$

$$\text{Total present value} = \underbrace{\frac{50}{1.04} + \frac{50}{(1.04)^2} + \cdots + \frac{50}{(1.04)^{10}}}_{\text{coupons}} + \underbrace{\frac{1000}{(1.04)^{10}}}_{\text{principal}}$$

$$= \frac{50}{1.04}\left(1 + \frac{1}{1.04} + \cdots + \frac{1}{(1.04)^9}\right) + \frac{1000}{(1.04)^{10}}$$

$$= \frac{50}{1.04}\left(\frac{1 - \left(\frac{1}{1.04}\right)^{10}}{1 - \frac{1}{1.04}}\right) + \frac{1000}{(1.04)^{10}}$$

$$= 405.545 + 675.564$$

$$= \$1081.11$$

17. (a) The people who receive the 100 billion dollars in tax rebates spend 80% of it, for an initial amount spent of $100(0.80) = 80$ billion dollars. The people who receive this 80 billion dollars spend 80% of it, for an additional amount spent of $80(0.80)$, and so on. We have

$$\text{Total amount spent} = 80 + 80(0.80) + 80(0.80)^2 + 80(0.80)^3 + \cdots.$$

Notice that the initial amount spent is not the original tax rebate of 100 billion dollars, but 80% of 100 billion dollars, or 80 billion dollars. The total amount spent is an infinite geometric series with $a = 80$ and $r = 0.80$. Since $-1 < r < 1$, the series converges and we have

$$\text{Total amount spent} = \frac{80}{1 - 0.80} = 400 \text{ billion dollars.}$$

(b) If everyone who receives money spends 90% of it, then the initial amount spent is $100(0.90) = 90$ billion dollars. The people who receive this 90 billion dollars spend 90% of it, and so on. We have

$$\text{Total amount spent} = 90 + 90(0.90) + 90(0.90)^2 + 90(0.90)^3 + \cdots.$$

This is an infinite geometric series with $a = 90$ and $r = 0.90$. Since $-1 < r < 1$, the series converges and we have

$$\text{Total amount spent} = \frac{90}{1 - 0.90} = 900 \text{ billion dollars.}$$

Notice that an increase in the spending rate from 80% to 90% causes a dramatic increase in the total effect on spending.

Solutions for Section 10.3

1. After receiving the nth injection, the quantity in the body is 50 mg from the nth injection, $50(0.60)$ from the injection the previous day, $50(0.60)^2$ from the injection two days before, and so on. The quantity remaining from the first injection (which has been in the body for $n - 1$ days) is $50(0.60)^{n-1}$. We have

$$\text{Quantity after } n^{\text{th}} \text{ injection} = 50 + 50(0.60) + 50(0.60)^2 + \cdots + 50(0.60)^{n-1}.$$

(a) The quantity of drug in the body after the 3rd injection is

$$\text{Quantity after } 3^{\text{rd}} \text{ injection} = 50 + 50(0.60) + 50(0.60)^2 = 98 \text{ mg.}$$

Alternately, we could find the sum using the formula for a finite geometric series with $a = 50$, $r = 0.60$, and $n = 3$:

$$\text{Quantity after } 3^{\text{rd}} \text{ injection} = \frac{50(1 - (0.60)^3)}{1 - 0.60} = 98 \text{ mg.}$$

(b) Similarly, we have

$$\text{Quantity after } 7^{\text{th}} \text{ injection} = 50 + 50(0.60) + 50(0.60)^2 + \cdots + 50(0.60)^6.$$

We use the formula for the sum of a finite geometric series with $a = 50$, $r = 0.60$, and $n = 7$:

$$\text{Quantity after } 7^{\text{th}} \text{ injection} = \frac{50(1 - (0.60)^7)}{1 - 0.60} = 121.5 \text{ mg.}$$

(c) The steady state quantity is the long-run quantity of drug in the body if injections are continued indefinitely. In the long run,

$$\text{Quantity right after injection} = 50 + 50(0.60) + 50(0.60)^2 + \cdots.$$

This is an infinite geometric series with $a = 50$ and $r = 0.60$. Since $-1 < r < 1$, the series converges. Its sum is

$$\text{Quantity right after injection} = \frac{50}{1 - 0.60} = 125 \text{ mg.}$$

5. (a) The steady state quantity is the quantity of drug in the body if tablets are taken daily for the long run. Right after a tablet is taken, in the long run we have

$$\text{Quantity after tablet} = 120 + 120(0.70) + 120(0.70)^2 + \cdots.$$

This is an infinite geometric series with $a = 120$ and $r = 0.70$. Since $-1 < r < 1$, the series converges and we use the formula for the sum of an infinite geometric series: In the long run,

$$\text{Quantity right after tablet} = \frac{120}{1 - 0.70} = 400 \text{ mg.}$$

(b) Right after a tablet is taken, at the steady state there are 400 mg of the drug in the body. In one day, 30% of this quantity, or $400(0.30) = 120$ mg, is excreted. This is equal to the quantity that is ingested in one tablet.

9. Since the toxin is metabolized at a continuous rate of 0.5% per day, the quantity remaining one day after consuming a single quantity of 8 micrograms is $8e^{-0.005}$. The person is consuming 8 micrograms every day, so the total accumulated toxin the person has in the body right after consuming the toxin is the sum of 8 (from the quantity just consumed) + $8(e^{-0.005})$ (from the quantity consumed the previous day) + $8(e^{-0.005})^2$ (from the quantity consumed two days ago), and so on. The total accumulation in the body is the sum of an infinite geometric series with $a = 8$ and $r = e^{-0.005}$. Since $-1 < r = 0.9950124 < 1$, the series converges to the sum:

$$\text{Total accumulation right after lunch} = 8 + 8(e^{-0.005}) + 8(e^{-0.005})^2 + \cdots$$
$$= \frac{8}{1 - e^{-0.005}} = 1604.0 \text{ micrograms.}$$

Since the person consumes 8 micrograms each day, the total accumulation of the toxin right before lunch is $1604 - 8 = 1596$ micrograms.

13. (a) In this case, consumption remains 3 trillion m^3 per year. Since reserves are 180 trillion m^3, the reserves are exhausted in n years, where

$$3n = 180$$
$$n = \frac{180}{3} = 60 \text{ years.}$$

(b) Assuming consumption increases each year by a factor of 1.05, consumption in 2008 is predicted to be $3(1.05)$ trillion m^3; in 2009, it is predicted to be $3(1.05)^2$; and in 2010, it is predicted to be $3(1.05)^3$, and so on. Representing total usage in n years by Q_n trillion m^3, we have

$$Q_n = 3(1.05) + 3(1.05)^2 + 3(1.05)^3 + \cdots + 3(1.05)^n.$$

Using the formula for the sum of a finite geometric series, we have

$$Q_n = 3(1.05)\frac{1 - (1.05)^n}{1 - 1.05} = 63((1.05)^n - 1).$$

To find how long reserves will last, we set $Q_n = 180$ and solve for n:

$$63((1.05)^n - 1) = 180$$
$$(1.05)^n - 1 = \frac{180}{63} = 2.857$$
$$(1.05)^n = 3.857$$
$$n \ln(1.05) = \ln(3.857)$$
$$n = \frac{\ln(3.857)}{\ln(1.05)} = 27.7 \text{ years.}$$

17. If usage decreases by 4% each year, then consumption of the mineral this year is 5000 m³, consumption next year is predicted to be $5000(0.96)$, consumption the following year is predicted to be $5000(0.96)^2$ and so on. Total consumption during the next n years is given by

$$\text{Total consumption for } n \text{ years } = 5000 + 5000(0.96) + 5000(0.96)^2 + \cdots + 5000(0.96)^{n-1}.$$

This is a finite geometric series with $a = 5000$ and $r = 0.96$. We have

$$\text{Total consumption for } n \text{ years } = \frac{5000(1 - (0.96)^n)}{1 - 0.96}.$$

If we try to find the value of n making total consumption equal to 350,000, we see that there are no such values of n. Why? If we consider consumption of this mineral forever under these circumstances, we have the infinite geometric series:

$$\text{Total consumption forever } = 5000 + 5000(0.96) + 5000(0.96)^2 + \cdots.$$

Since $-1 < r = 0.96 < 1$, this infinite series converges and we have

$$\text{Total consumption } = \frac{5000}{1 - 0.96} = 125{,}000 \text{ m}^3.$$

If usage of this mineral decreases by 4% per year, we can use the mineral forever and the total reserve never runs out.

Solutions for Chapter 10 Review

1. The sum can be rewritten as

$$2 + 2(2) + 2(2^2) + \cdots + 2(2^9).$$

This is a finite geometric series with $a = 2$, $r = 2$, and $n = 10$. We have

$$\text{Sum } = \frac{2(1 - (2)^{10})}{1 - 2} = 2046.$$

5. This is an infinite geometric series with $a = 30$ and $r = 0.85$. Since $-1 < r < 1$, the series converges and we have

$$\text{Sum } = \frac{30}{1 - 0.85} = 200.$$

9. (a) (i) On the night of December 31, 1999:

First deposit will have grown to $2(1.04)^7$ million dollars.
Second deposit will have grown to $2(1.04)^6$ million dollars.
. . .

Most recent deposit (Jan.1, 1999) will have grown to $2(1.04)$ million dollars.

Thus

$$\text{Total amount} = 2(1.04)^7 + 2(1.04)^6 + \cdots + 2(1.04)$$
$$= 2(1.04)\underbrace{(1 + 1.04 + \cdots + (1.04)^6)}_{\text{finite geometric series}}$$
$$= 2(1.04)\left(\frac{1 - (1.04)^7}{1 - 1.04}\right)$$
$$= 16.43 \text{ million dollars.}$$

(ii) Notice that if 10 payments were made, there are 9 years between the first and the last. On the day of the last payment:

First deposit will have grown to $2(1.04)^9$ million dollars.
Second deposit will have grown to $2(1.04)^8$ million dollars.
$\cdots$

Last deposit will be 2 million dollars.

Therefore

$$\text{Total amount} = 2(1.04)^9 + 2(1.04)^8 + \cdots + 2$$
$$= 2\underbrace{(1 + 1.04 + (1.04)^2 + \cdots + (1.04)^9)}_{\text{finite geometric series}}$$
$$= 2\left(\frac{1 - (1.04)^{10}}{1 - 1.04}\right)$$
$$= 24.01 \text{ million dollars.}$$

(b) In part (a) (ii) we found the future value of the contract 9 years in the future. Thus

$$\text{Present Value} = \frac{24.01}{(1.04)^9} = 16.87 \text{ million dollars.}$$

Alternatively, we can calculate the present value of each of the payments separately:

$$\text{Present Value} = 2 + \frac{2}{1.04} + \frac{2}{(1.04)^2} + \cdots + \frac{2}{(1.04)^9}$$
$$= 2\left(\frac{1 - (1/1.04)^{10}}{1 - 1/1.04}\right) = 16.87 \text{ million dollars.}$$

Notice that the present value of the contract ($16.87 million) is considerably less than the face value of the contract, $20 million.

13. The amount of additional income generated directly by people spending their extra money is $100(0.8) = \$80$ million. This additional money in turn is spent, generating another $(\$100(0.8))(0.8) = \$100(0.8)^2$ million. This continues indefinitely, resulting in

$$\text{Total additional income} = 100(0.8) + 100(0.8)^2 + 100(0.8)^3 + \cdots = \frac{100(0.8)}{1 - 0.8} = \$400 \text{ million}$$

17. (a) The quantity of atenolol in the blood is given by $Q(t) = Q_0 e^{-kt}$, where $Q_0 = Q(0)$ and k is a constant. Since the half-life is 6.3 hours,

$$\frac{1}{2} = e^{-6.3k}, \quad k = -\frac{1}{6.3}\ln\frac{1}{2} \approx 0.11.$$

After 24 hours

$$Q = Q_0 e^{-k(24)} \approx Q_0 e^{-0.11(24)} \approx Q_0(0.07).$$

Thus, the percentage of the atenolol that remains after 24 hours $\approx 7\%$.

(b)

$$Q_0 = 50$$
$$Q_1 = 50 + 50(0.07)$$
$$Q_2 = 50 + 50(0.07) + 50(0.07)^2$$
$$Q_3 = 50 + 50(0.07) + 50(0.07)^2 + 50(0.07)^3$$
$$\vdots$$
$$Q_n = 50 + 50(0.07) + 50(0.07)^2 + \cdots + 50(0.07)^n = \frac{50(1 - (0.07)^{n+1})}{1 - 0.07}$$

(c)

$$P_1 = 50(0.07)$$
$$P_2 = 50(0.07) + 50(0.07)^2$$
$$P_3 = 50(0.07) + 50(0.07)^2 + 50(0.07)^3$$
$$P_4 = 50(0.07) + 50(0.07)^2 + 50(0.07)^3 + 50(0.07)^4$$
$$\vdots$$
$$P_n = 50(0.07) + 50(0.07)^2 + 50(0.07)^3 + \cdots + 50(0.07)^n$$
$$= 50(0.07) \left(1 + (0.07) + (0.07)^2 + \cdots + (0.07)^{n-1} \right) = \frac{0.07(50)(1 - (0.07)^n)}{1 - 0.07}$$

STRENGTHEN YOUR UNDERSTANDING

1. True, since there is a constant ratio of 2 between successive terms.

5. True, since this is a geometric series with 11 terms, with first term 1 and constant ratio $1/3$.

9. True. The first term is $1/3$, so the sum is $(1/3)/(1 - (1/3)) = (1/3)/(2/3) = 1/2$.

13. True, since without any interest, we would need to deposit $6000 \cdot 10 = 60{,}000$ dollars in the annuity today to pay 6000 dollars for 10 years. With the addition of interest, we can deposit less than \$60,000.

17. True. The account will earn $735{,}000(0.05) = 36{,}750$ dollars a year so it can generate \$35,000 annual payments in perpetuity.

21. False. Since the person is metabolizing the drug throughout the day, the person will have less than the full 100 mg in the body.

25. False. The level goes up right after each new dose.

29. False; the correct value for long-term quantity is $50/(1 - e^{-0.05})$.

CPSIA information can be obtained at www.ICGtesting.com
Printed in the USA
BVOW09s0623200315

392499BV00007B/8/P